T0074739

*Praise for Lucy Jones's*

# LOSING EDEN

"Fascinating. The connection between mental health and the natural world turns out to be strong and deep—which is good news in that it offers those feeling soul-sick the possibility that falling in love with the world around them might be remarkably helpful."     —Bill McKibben

"[Jones] has more than done her homework. . . . Her book, filled with examples of nature's benefits to human health, makes a strong argument for rescuing nature before it's too late."     —*Booklist*

"In meticulous detail, Jones quests to bring us an impressive array of answers to the question of whether 'nature connection' has a tangible effect on our minds. . . . Her results are compelling. . . . This book will convince you that nature is an intrinsic part of ourselves."

—*The Irish Times*

"Fascinating. . . . Written in such lush, vivid prose that reading it . . . one can feel transported and restored."     —*The New Statesman* (London)

"Urgent. . . . Accessible, moving. . . . A beautifully written, research-heavy study about how nature offers us well-being."     —*The Observer*

"Compelling. . . . The book is not really a memoir; it's about all of us."
—*The Times Literary Supplement* (London)

"A passionate and thorough exploration of the growing scientific evidence showing why humans require other species to stay well."

—*The Guardian*

Lucy Jones

# LOSING EDEN

Lucy Jones was born in Cambridge, England, and educated at University College London. She has written extensively on culture, science, and nature. Her articles have been published on BBC Earth and in *The Sunday Times*, *The Guardian*, and *The New Statesman*. Her first book, *Foxes Unearthed*, received the Society of Authors' Roger Deakin Award. Jones lives in Hampshire, England.

lucyfjones.com

# LOSING EDEN

# LOSING EDEN

Our Fundamental Need for the Natural World
and Its Ability to Heal Body and Soul

## LUCY JONES

VINTAGE BOOKS

A DIVISION OF PENGUIN RANDOM HOUSE LLC

NEW YORK

The Library of Congress has cataloged the Pantheon edition as follows:
Names: Jones, Lucy (Journalist), author.
Title: Losing Eden : our fundamental need for the natural world
and its ability to heal body and soul / Lucy Jones.
Description: First United States edition. | New York : Pantheon Books, 2021. |
Includes bibliographical references and index.
Identifiers: LCCN 2020045787 (print) | LCCN 2020045788 (ebook)
Subjects: LCSH: Human beings—Effect of environment on. | Nature—Social
aspects. | Nature—Psychological aspects. | Well-being.
Classification: LCC GF51 .J66 2021 (print) | LCC GF51 (ebook) |
DDC 304.2—dc23
LC record available at https://lccn.loc.gov/2020045787
LC ebook record available at https://lccn.loc.gov/2020045788

**Vintage Books Trade Paperback ISBN: 978-0-593-08295-9**
**eBook ISBN: 978-1-5247-4933-0**

*Author photograph © Rupert Van den Broek*
*Book design by Cassandra J. Pappas*

vintagebooks.com

*For my mother and father*

# Contents

## PART VI · SNAG

# Prologue

Xena walked down the street towards her grandmother's house. It was a boiling hot day, but she'd remembered her hat, respirator and sun goggles. She walked as fast as she could down the pavement, then through the concrete tunnel and up the covered stone stairway to get out of the blaze of the sun. She could hear the boom of the high-speed train in the next borough as she walked. She crossed the road to walk over to the pop-up green space—Astroturf—but thought better of it: recently, it had got so hot that the plastic grass had melted onto her friend's sandals. Xena took the longer route. Even the artificial trees couldn't provide shelter from the heat today. The mountains in the distance were almost blotted out by the smoke of the wildfires, and she could barely see past twenty metres ahead. All was grey. A bus passed with an advert for a new telegarden programme, scheduled for 2102. People could log in via their brain chip, and plant and water virtual seeds to see them grow. She made a mental note to mention it to her granny.

Granny couldn't often leave the house any more, so Xena would have to go and visit her, but she didn't mind. She had a holographic nature scene (HNS) set up in the living room and Xena always felt happier and less stressed-out after a visit. The HNS she liked best had actual trees on one side, which were a kind of greeny-brown. In the middle of the screen was a lake, and occasionally she'd see a fish leap out of the water. The lake looked clean, nothing like the dirty, stinking basins and rivers near Xena's home. Her favourite thing about the HNS was the sound. It was a kind of music she'd never heard before: birds singing, frogs croaking, something barking. She'd seen birds in the local museum and her school sometimes piped birdsong into the classroom, but she'd never seen one in real life. She wondered if Granny had.

Xena arrived and rang the doorbell. Her breathing began to slow, though a slight rasp remained, and she wiped the sweat from her forehead. After a minute or so, Granny opened the door and beckoned her in. She stroked her head, squeezed her hand and led her through the apartment.

Xena was relieved to see the HNS was working and she climbed onto the sofa and curled up.

"I've got a new one for you, darling," said Granny.

She drew a letter "H" over her implant and the hologram appeared. At first the scene was foggy and it was hard to see anything, but when the mist cleared Xena saw a group of very tall trees with all sorts of limbs and bits coming off them. Then she noticed something small and bright green. Suddenly, it leaped into the air and disappeared.

"What was that, Granny?"

"Oh, that's a . . . tree frog. This is a tropical rainforest."

"Tropical rainforest," repeated Xena slowly.

Three birds—well, she thought they were birds—flew through the scene. They had long orange noses and black and white bodies. She couldn't believe that they could stay up in the air with such long noses. She followed the birds and her eyes rested on a tiny creature with big yellow eyes nestled in a branch.

"What is that, Granny?" she squealed.

"An owl, darling, maybe a baby owl."

"This is the best HNS so far, Granny," she said.

"I wish you could've seen it in real life."

"Birds in real life, like every day?"

"Yes, and other animals too."

"Actually walking around? Not in a zoo?"

"Sometimes. And insects. Do you know about butterflies?"

"We were told about them at school."

"England was full of butterflies. You could sit in a garden or park and spot quite a few different species in the summer."

"What was it like, Granny?"

"Oh, it was . . ."

Granny paused. Xena looked around. Alarmed, she saw that Granny seemed to be crying.

"Granny!"

Granny cleared her throat.

"Was it like this, but in real life?" She gestured to the hologram.

"Well, yes, if you were in the rainforest," said Granny. "In England, in my garden, you'd see small creatures called bumblebees that looked like tiny bears, in black and yellow. In the warmer months, you could hear the hum of insects looking for nectar. My favourite butterfly had black stripes over its orange wings, so it looked a bit like a flying tiger. There were trees called oaks, which lived for hundreds of years. The garden looked different every day."

"Could you touch the trees, Granny?"

"Oh yes. You could touch the leaves and the plants and the flowers."

"What did they feel like?"

"Soft, I suppose, but each one felt different. Dandelions were called a weed, but in the early summer they turned into these perfect bulbous globes which you could blow and all the furry seed heads would fill the air."

"Like magic?"

"Yes, in a way. We called them 'dandelion clocks.' And the scents were so good. Each flower smelled different. I loved the smell of roses, the bluebells, the pine trees . . . Oh, do you know about conkers?"

"No, what's that?"

"In the spring—that was the season when everything came out in bloom—the horse chestnut tree did this thing called candling, where it would grow flowers that looked like ice-cream sundaes. Later, the plant would produce these bright green balls covered in spikes. When they dropped off the tree, we'd look inside their cracked shells to find something we called conkers. They were brown and shiny and meant that autumn—that was another season—was here and the leaves would soon change colours, from green to red or orange or yellow."

"You're kidding, Granny!"

Granny shook her head.

"And you saw that every day if you wanted to?"

"Yes, darling."

"What was it like?"

"It was . . . wonderful."

"Why did nature end, Granny?"

Granny sighed. "We didn't love it enough," she said. "And we forgot that it could give us peace."

PART I

# SEEDLING

# Introduction

## The Baby in the Soil

Do you not see the very shrubs, the sparrows, the ants, the spiders, and the bees, all busied, and in their several stations cooperating to adorn the system of the universe?    —MARCUS AURELIUS, *Meditations*

Every leaf speaks bliss to me.    —EMILY BRONTË, "Fall, Leaves, Fall"

The din of the dusty world and the locked-in-ness of human habitations are what human nature habitually abhors; while, on the contrary, haze, mist and the haunting spirits of the mountains are what human nature seeks, and yet can rarely find.    —GUO XI, eleventh century

O NE LATE-SUMMER AFTERNOON, I was sitting at the edge of a wildflower bed in my garden with my baby daughter, touching desiccated seed heads and looking for worms in the soil. Spiders the colour of syrup had sprung up in webs across the garden, their coracle rumps jewel-like in the sun. Although it was August, it already felt like autumn in southern England. The apples and plums were falling, the ground was jam-slicked and wasp-studded, and the leaves were turning. As I pointed out where the hedgehog came out for beetles and caterpillars at night, I looked at my daughter, and felt a chill.

The papers were full of droughts, floods, extreme weather events and high temperatures, sometimes even beyond what scientists had predicted. What was coming for her and her generation? Climate chaos was accelerating. Ice was melting faster than expected. The world seemed to be on fire. Closer to home, the seasons felt off—autumn in August, midwinter in March. Every day brought news of another species in fast decline. Swifts, swallows, hedgehogs, all were on the road to extinction. Would there be any ancient woodlands left or old oaks to climb and wonder at? How many more species of bird would join the Spix's macaw, the poo-uli, the Pernambuco pygmy-owl and the cryptic tree-hunter and die out this century? With 80 percent of Europe and the United States already without their dark skies because of light pollution, would she ever see the Milky Way? And what would this "biological annihilation," as scientists had put it, do to her mind and spirit, assuming she managed to survive at all?

Around that time, I read about a depressing concept, coined by the American author, ecologist and lepidopterist Robert Pyle: the "extinction of experience." As fewer children connect with nature, it will follow, he argues, that if they become parents, their children will in turn have an even more tenuous connection with the natural world. "Its premise involves a cycle of disaffection and loss that begins with the extinction of hitherto common species, events, and flavours in our own immediate surrounds; this loss leads to ignorance of variety and nuance, thence to alienation, apathy, an absence of caring, and ultimately to further extinction."

I could see this pattern in my own family. My grandmothers had an inherent lexicon of the natural world and how it operates. My parents knew their birds, flowers and plants; names, timing and behaviours. I knew a bit, maybe 5 to 10 percent of what they knew, and I was keener on wildlife than most of my friends. It would follow that my daughter's connection with the natural world would be even more remote than mine. Would she be able to name—by which I mean know—anything at all? Or would she be so desensitized to the point where a connection with nature would have little, or no, value? As Pyle wrote, "What is the extinction of the condor to a child who's never known a wren?"

We have never been at this point of disconnection with the rest of nature before. In Britain, half of our ancient woodland has disappeared in the last eighty years. During the twentieth century, 97 percent of lowland meadows and 90 percent of coppiced woodland in England and Wales was lost, along with the animal and plant communities that lived in them. More than one in ten species are now under threat of extinction in the United Kingdom. Over just the last fifty years the populations of mammals, birds, reptiles and fish have fallen by 60 percent worldwide.

Our behaviour has changed as the landscape has winnowed. Simply put, we've moved inside. We live in cubicles, cars and tower blocks, spending only 1 to 5 percent of our time outdoors. We're used to surviving outside the rhythms of the natural world. Our need, opportunities and desire to interact with the rest of nature have dramatically decreased.

In 2005, the influential American writer Richard Louv coined the phrase "nature-deficit disorder" to refer to the impact of a lack of connection with nature on people's health. "It describes the human costs of alienation from nature, among them: diminished use of the senses, attention difficulties, and higher rates of physical and emotional illnesses," he wrote. Since then, the disconnection is starting to make its way into our language. In the same decade, the Australian philosopher Glenn Albrecht, frustrated that there were few concepts in the English language to help make sense of the relationship between humans, the built environment, the natural environment and our psychological states, coined the word "psychoterratic," which describes both earth- (terra) and mental health (psyche)–related emotions, feelings and conditions. Psychoterratic illnesses, for example, are earth-related mental health issues such as ecoanxiety and global dread. "Solastalgia"—an admixture of solace, nostalgia and destruction—describes a feeling of nostalgia and powerlessness about a place that once brought solace which has been destroyed. Another new term is "species loneliness," to mean a collective sorrow and anxiety arising from our disconnection from other species. The environmental writer Robin Wall Kimmerer describes it as "a deep, unnamed sadness stemming from estrangement from the rest of Creation, from the loss of relationship."

And yet, judging by the way we treat forests and fens, the seas and the rivers, and the wildlife that live in them, it would appear that industrialized society perceives nature to be little more than a nicety: a luxury, an extra, a garnish—"Green crap," as the former prime minister David Cameron reportedly called environmental policies—rather than the support system that keeps us all alive.

Certainly my daughter and her generation were born into a time of extraordinary disconnection, rapid climate destruction and psychological retreat from the rest of the living world. And yet, they are also connected to nature through story and language that stretch back into the deep past.

Humans have long turned to aspects of the natural world—particularly animals, landscapes, weather patterns and biological processes—as a way to interpret and make sense of our existence. From simple, practical idioms—"out on a limb," "the early bird catches the worm," "down to earth"—to vast cosmic symbols of renewal, regeneration and tenacity, nature imagery helps us to understand and extract meaning from the world we find ourselves in. Of course our earliest creation myths and cosmologies are filled with shared motifs from the natural world—floods, serpents, eggs and animistic beliefs—as early humans were much closer to nature. But, despite our disconnection, we still turn to it today. We even reach for it when we think about the internet: "web," "stream," "raw data." We are deeply entwined with the rest of nature on a linguistic and mental level; we have created our language, culture and consciousness—the most essential parts of human psychology, from which our desires and preferences flow—within, and in close relationship to, the natural environment we have lived in for millennia. The writer and naturalist Richard Mabey puts this well: "Our imaginative affinities with the natural world are a crucial *ecological* bond, as essential to us as our material needs for air and water and photosynthesising plants."

Our aesthetic preference for nature's flora and fauna can be traced through human history to the modern day. Ancient urban communities in Byzantium, Persia and medieval China were designed with ornamental gardens; Pompeii was decorated with frescos of natural scenes; Cistercian monks grew flowers and planted trees for the sake of their beauty from the twelfth century. We have also brought the natural world inside with us, from the paintings of animals on the walls of caves and woodblock prints of flowers, mountains, storms and waves in China and Japan, to the European tradition of bringing an evergreen tree into the house each winter. These days, the most popular PC or laptop wallpapers are images of cherry blossom, autumn leaves and turquoise seas; millennials, in particular, bring potted plants into their flats and houses in increasing numbers, and in 2017, Pantone's colour of the year was Spring Green. (The company decides on the colour of the year by reading the "mood and attitude" of the global culture. Green was chosen because it "symbolizes the reconnection we seek with nature.")

So why exactly are we drawn to these elements?

BEYOND THE PHILOSOPHICAL or aesthetic, we have an even more urgent, even more basic need for nature—our very well-being depends on it. The association between good health and a connection to, or experience of, a healthy and beautiful natural environment has been entrenched throughout our long history, and much discussed and written about by our ancestors. The ancient Sumerian myth of Enki described Dilmun, the garden paradise said to be the inspiration for the Garden of Eden, as a place where "human beings are untouched by illness." Early Sanskrit literature makes the emotional connection too: "There are none happy in the world but beings who enjoy freely a vast horizon," said Damodara (Krishna). Virgil's account of Arcadia, from his *Eclogues* (also known as the *Bucolics*), was a landscape of comfort and healing—with its cool springs, zephyrs, laurels and tamarisks—for Gallus, who was dying of a broken heart.

From the creation of urban parks for civic health to the garden city movement, for centuries people have often acted on an intuitive sense that human beings require communion with the natural world for their emotional selves, for their nerves and their psyches. Now modern science is catching up.

Our need for nature as a healing force became especially acute and explicit as the West became industrialized and our contact and connection with the natural world around us dwindled. As people moved to cities, and away from the land, they had to actively seek out nature as they became physically separated from it. In the pre-industrialized West, the wilderness was often seen as cruel, revolting and ugly: when the Italian poet Petrarch climbed Mont Ventoux in 1336, he chastised himself for "admir[ing] earthly things" and fled angrily from the peaks in shame.

But from the eighteenth century onwards people began to see the natural landscape in a new way, and with the rise of the Romantic movement in art and poetry, and Transcendentalists such as Henry David Thoreau and Ralph Waldo Emerson, a new age of emotional sensibility relating to place and landscape was ushered in. The rise of the middle classes saw an increase in travel across Britain to enjoy the scenery, following Wordsworth in seeking spiritual experiences in the rugged hills of the Lake District, and viewing mountains and hills that had previously been seen as pimples, warts or blisters on the surface of God's earth—so revolting that on some trains the blinds were pulled down to avoid offence. One peak was called the "Divels-Arse." Nature became a legitimate preoccupation of painters and composers, as well as of poets and philosophers, throughout Europe. As the contemporary environmental historian and California-based academic Roderick Nash explains, appreciation of wilderness began in the cities and the "civilizing process which imperils wild nature is precisely that which creates the need for it."

Post-industrialization, natural places started to be used as venues to treat people in emotional and psychic pain. In England in 1796, the York Retreat was developed in the countryside by the Quaker community. Patients living in lunatic asylums were encouraged to walk in the gardens and spend time with domestic animals such as poultry, as well as with birds and flowers. The 1827 Code of Practice for lunatic asylums in Britain stated that courtyards should be airy, and offer "some prospect over the walls." The Kirkbride mental hospitals, built across the United States in the mid to late nineteenth century, had at least a hundred acres of land and the design requirement that green space should be "fertile" and "exhibit life in its active forms."

In 1859, Florence Nightingale wrote about the power of nature in aiding recovery. In *Notes on Nursing*, she reported her observations on the effect of viewing nature on her patients. "I have seen, in fevers (and felt, when I was a fever patient myself), the most acute suffering produced from the patient (in a hut) not being able to see out of [the] window, and the knots in the wood being the only view. I shall never forget the rapture of fever patients over a bunch of bright-coloured flowers," she recorded. "I remember (in my own case) a nosegay of wildflowers being sent me, and from that moment recovery becoming more rapid." She observed: "Little as we know about the way in which we are affected by form, by colour, and light, we do know this, that they have an actual physical effect. Variety of form and brilliance of colour in the objects presented to patients are actual means of recovery."

Today, there is a growing field of "nature therapy" and a mounting evidence base to show why and how a connection with the rest of nature is good for our minds. Perhaps we are noticing this all the more now, as we are in danger of losing the living world as we have known it, and with it, potentially, part of ourselves.

Some argue the "extinction of experience" matters because it has led and will continue to lead to widespread public indifference to the loss of our natural environment, and therefore makes tackling future environmental catastrophe harder. Others point out the global increase in myopia or near-sightedness in young people, with lack of time spent outdoors cited as a factor. Others worry about the "pandemic" levels of vitamin D deficiency, which is more common in children who spend over four hours at computer screens indoors, and the increase in rates of rickets, a symptom of the deficiency. Some argue it matters because children are facing an obesity epidemic and sitting indoors instead of playing outside makes it worse. Some point to the mental health crisis in the West and the rising number of "deaths of despair," that is, deaths resulting from alcoholism (liver disease), drug addiction (poisoning) and depression (suicide), that are causing life expectancy in the United States to fall. People are increasingly dying for psychiatric, social and existential reasons. In the United States, the suicide rate has increased by 25 percent since 1999: experts point to the growing suicide rates among teenagers in Britain as evidence of a mental health crisis.

As I sat there in the late August sun, watching my daughter gently touch and explore the white petals and yellow eyes of the final daisies of the year, smelling the last of the sweet clover in our tiny wildflower patch, I was concerned that our exclusion from nature could lead—or has indeed led—to a kind of psychic crisis or, at the least, the absence of something profoundly important for human psychological health. But there seemed little point just getting melancholic about it. Rather, I decided to set out to see how, why and by which mechanism a relationship with the natural world—or lack of one—can affect human mental health, at all stages of life.

I HAVE TO CONFESS a personal interest, as a connection to the natural world helped me at my lowest point. Not that that connection had always been there: when I was growing up in the Thames Valley in the late 1990s and early 2000s, nature wasn't very "cool," and I quickly became more interested in Radiohead, customizing my flares and reading the Beat poets than in the wildlife I was obsessed with as a child. I rarely chose to spend time directly in the natural environment, beyond smoking in bushes or behind trees.

More of my time was spent drinking. I discovered at the age of fourteen that alcohol could make me feel comfortable in my own skin. It set me up in a kind of eyrie. From the off, I enjoyed its transformative effect, and the way it hurried me out of my head: *glug, glug, click: SSHHhhhh.* It turned off an internal monologue that could be self-lacerating and acid-tongued. At first, drinking was mostly a hoot, even if I usually felt a little uneasy and ashamed the next day. I seemed to take it further than most of my friends. I drank quicker than other people—and more, and longer. I downed drinks for a swifter buzz. I started to drink on my own. Often I'd vomit, but that never made me call it a night.

Adolescence was also when I experienced mental ill health for the first time. When I was around seventeen, something snapped. Half-term arrived and I had a run of fun things planned every day and every night. It was hot and the streets of London smelled of sweat, exhaust fumes and cigarette smoke. And then, suddenly, all the colour was sucked out. I didn't feel joy or happiness about anything. Food lost its flavour. Getting out of bed took significant effort. Talking was tiring, so instead I slept. I wanted to be unconscious. I was frightened by how numb and unalive I felt, not knowing what was happening to me. I'd felt sad before, of course, and had bad days, but not to feel anything positive was disturbing. I started to cancel my plans. To counteract the dull ache—it was as if I had a flat battery—I turned, with renewed intention, to drink.

Early on, booze had felt like a nest, a safe place, a warm blanket of protection that meant I could live and talk the way I wanted to without second-guessing my self-conscious self. At first, it had been a kind of magic potion that made it easier to connect with people, to talk in public, to feel less shy. It had become a kind of adolescent scaffolding. I didn't realize I was beginning, slowly, to lose myself. I didn't realize that the endpoint for an addict—especially when you add cocaine to the mix—isn't connection, but loneliness, obsession and isolation.

I didn't imagine that I might end up, a decade later, trying to nap on the floor of a bathroom cubicle in the office during my lunch breaks. Curled up in the foetal position, the wood of the door feels calming, beige, while the strong light is hurting my eyes. The floor is cold and hard like marble, soothing a hot, angry, pulsing temple. The vein is trying to worm its way out of the skin. Breathing deeply to avoid throwing up again, I start to nod off. An alarm sounding and snoozing, repeatedly. The drifting voices of other colleagues coming in and out of the bathroom—I wonder, have they clocked the noise of the alarm, do they know I'm in here to sleep off a raging hangover? Fuck, I'm going to … Waiting for silence, washing my face, wiping bloodshot, red-rimmed eyes with shaking hands, putting a piece of chewing gum in my mouth, taking a deep breath and returning to my desk. And then, eventually, seeing that this is a regular occurrence, realizing that I can walk out of the locked cubicle, but not away from this pattern—and it's starting to get nasty; the flames are licking at my heels. When I try to leave, I find myself trapped: the eyrie, once safe, has become treacherous.

FOUR ELEMENTS WERE crucial in my recovery. Three were straight-forward: psychiatry and medicine, psychotherapy and the support of friends, family and other addicts. The fourth, I didn't quite understand, which is the genesis of this book.

I knew that the drink and drugs had to go: the highs were getting harder to achieve; the lows were becoming more dangerous and self-destructive. My mental and physical health were both deteriorating, but I couldn't seem to manage to stop drinking or using drugs on my own. So, after months of trying, I found a rehabilitation programme and started to collect days without a drink or other substance. That first year, I felt as if I was walking around unpeeled. Everything was heightened and intense. I was a soft-shell young adult again, for I had been drinking away my emotional life since early adolescence. The hard stuff was psychological. I had to learn to sit with new feelings. My fear about a sober life was that I'd never be at ease again. But after the shaky stage passed, I began to look up and out.

For support, I met up with other addicts, often in rooms in churches. Churches often have gardens, and gardens often have flowers. I noticed the colours, the pink and yellow of the petals on a bush near one of these rooms. It lit me up. I raised my hand and touched the soft petals with my fingers. I liked how the pink and yellow looked next to each other, like Battenberg cake. I started walking—drawn to trees, birds, flowers and plants with an urgency I'd only ever had for drugs, music or people.

Walking daily on Walthamstow Marshes, a large, open expanse next to a canal, to see kestrels, the shaggy old heron and the golden buttons of tansy made me feel safe and secure. It became my rehab: it soothed my rawness and patched me back together. The walks were both comforting—I would return to my flat calm of mind and light of spirit—and also exciting because of the wildlife and variety and colours I might encounter. The sky was different every day. The breeze on my face and the smell of the land and the touch of bark made me desirous of life. The state of active addiction is alienating. It involves a level of obsession with self and substance which can lead to isolation from others, even if you are regularly socializing. Nature had felt too wholesome for my sense of shame. My world had narrowed rapidly and I barely saw daylight: it was a vampirical existence and towards the end I cared about very little.

But in recovery, I connected with a lost, dormant part of myself, and I found this mirrored in the newfound sense of connectedness and openness I encountered in meeting with other addicts three or four times a week in those early years. I felt the glass pane between myself and other people lift; I felt newly connected to my friends and family—and to both society and the wider outdoor world. I was never lonely or alone on the marshes. I started to feel that I belonged to a wider family of species, a communion of beings, the matrix of life, from the spiders to the lichen and the cormorants to the coots. I felt born again. Nature picked me up by the scruff of my neck, and I rested in her teeth for a while.

The world glowed, and it cooled my mind. Nature softened the edges and sharp angles and stroked my hair and held my hand. I was caught, a burr on the leg of her, hooked on wonder and abundance.

At times, it was like a heady love affair, replete with obsession and awe and joy. At others, I simply turned up to the land, the river, the ponds for my spirit to be fed. I would leave feeling full up and satisfied, in a way that I never did with alcohol or cocaine. Paying attention to what was around me quietened the voices in my head. I tuned in to the beating wings of a skein of geese, a creaking cricket or the plop of a vole.

I had been sober for about a year when a pear tree outside my bedroom window in a flat in Hackney, East London, was concealed for six months by thick, ugly scaffolding. It was a beautiful tree—incandescent in spring, glowing green like kryptonite. The branches would bud and then they'd burst into curds of blossom, short-lived like a Perseid. The tree's trunk carried water and sap into the vessels and capillaries, up through the marrow, into the leaves. Its green deepened for the summer before becoming tumescent with fruit, blowing up the pears until they dropped to the ground—through heaviness and insect wear and tear—gloopy and fleshy. Then the leaves would dry, brown and drop quietly like divers as the tree brittled for winter, a still, hard mantis, belying unseen electrical messages and signals zig-zagging inside the bark and into the roots and out of the roots and into the trunk and back to the branches, the mycorrhiza fizzing under the ground. When the late winter days lost their tang and the world stretched and yawned, I would check back each morning for *tap, tap, creak,* new nubs soon to crack open, push forward and begin again.

This flat was a new start for me, a clean place with no memories of long binges stretching into frightening mornings—the mental terror of sobering up, and the resultant come-down—and the tree called me to keep going. It filled the sash window that I slept next to and sometimes worked beside, and I loved to watch its changing raiment and activity. Look how I grow and move and continue, it said. See how things change. I found a kind of emotional stability in its routine and fastened my hope to it. I'd sometimes find myself lying in bed on a Saturday afternoon, watching the gentle way it moved and how the sunlight danced on its leaves, in a trance. Afterwards, I'd feel relaxed and calm. I loved the tree and the way it made me feel. I often saw exquisitely coloured blue-tits, and once a woodpecker, though the birds were really the supporting cast. The flat was one of four in a house on a residential street a ten-minute walk from the nearest park. Only the people renting the basement flat had outdoor space—a small square of concrete—but we could climb out of the kitchen window and onto the ground-floor roof. It looked out to a housing estate, across a rank of poplars, and it was a fine place to read, chat to friends and listen to the trees. Compared with the place that I'd lived in before—on a nearby busy roundabout, above a hardware store—it felt much closer to nature, particularly the proximity to the pear tree. We hadn't consciously chosen the apartment for this reason—my only requirement then had been a bathtub—but it was a happy surprise. And it became something I grew to rely on.

I only understood quite how much when, after six months or so, the scaffolding for the upstairs neighbours' building work went up and the tree was blocked from view. What happened next made me realize the extent of my psychological need for the natural world. Within days, I felt a rising tension. I tried to peer around the lattice of metal bars to glimpse its vital green, to see how it was getting on, as if it were a drink that could quench my thirst, or a beloved I was separated from. I became frustrated that I could barely see it, irritable and morose. When I was at home, I felt caged. I sent passive-aggressive text messages to the neighbours to ask how long the scaffolding would be up for, and snapped at my then boyfriend.

Of course, it would have been extremely unusual if my neighbours had warned us that the pear tree would be concealed. Had the council decided it needed to be chopped down, we would have expected to have been notified in case of noise, or dust, but not because of any potential negative psychological side effects.

I didn't get over it. I felt unspooled for the six months or so it was hidden. Nor did I understand what was going on. I was embarrassed by the pricking of hot, furious tears and throat-tightening that would sometimes come on in the evening. I was freaked out by my emotional reaction. Could a tree—or the lack of one—really have such a strong psychological impact? What on earth was happening to me?

You see, this relationship took me by surprise. I hadn't been prescribed "nature," or advised to spend time outdoors. I'd more or less stumbled upon it. But I found I needed the natural world and used it in a similar way to my use of drugs and alcohol. I yearned for its comfort in the way that alcohol enveloped me. An added bonus: it didn't give me a hangover. Beforehand, I knew that walking by the sea or in the woods was a pleasant thing to do; indeed, that it could make people feel calm and relaxed. I vaguely knew that exercise or going for a walk with friends was good for one's mental health. I sometimes did that even when I felt low. But I didn't realize that the essence of it—the geometry, the scents, the sounds, the colours, the textures—could have such life-changing power. I didn't appreciate that a relationship with the rest of nature could help when I was mentally unwell.

What was behind this effect and why didn't I know about it before? Looking beneath the hood of what American landscape architect (and designer of Central Park in New York) Frederick Law Olmsted saw as "'Scenery' working by an unconscious process to produce relaxing and 'unbending' of faculties made tense by strain," I began to discover a plethora of scientific evidence, as well as stories and anecdotes, to explain what happens to the body and mind during that simple uplifting of the spirit.

Spurred on by my personal experience, and a growing sense, from the conversations I was having with scientists for my work as a journalist, that this was an exciting and important new area of research, I set out on a journey to explore the mechanisms by which contact with the natural world could affect the human mind.

At first, the question at the forefront of my research was relatively simple: How and why does nature make us feel good? How does it affect human mental health? And how can it ameliorate psychological pain? This took me to conferences in Britain and Germany to hear the latest cutting-edge research, to interviews with academics at the forefront of their fields, from neuroscientists in California to microbiologists in Eastern Europe. Where once the evidence that nature is needed for good human mental health was purely qualitative, now the quantitative research was mounting—and it was much more compelling, wideranging and varied than I'd anticipated. I had read many headlines and studies about the benefits on stress levels of walking in a park, and how gardening and outdoor swimming and spending time with birds and plants could be therapeutic, but I really wanted to understand why and how. At the outset, I hadn't realized that I was stepping into a relatively new, very fecund and exciting field of study. I hadn't known that many scientists, from those who study microbes to those who study the science of awe, were attempting—and succeeding—at proving categorically that nature has a measurable effect on human mental health. This book is partly an account of the most intriguing and compelling research that I found. There was no silver-bullet piece of evidence: I found the field of research to be as complex and diverse as an ecosystem, or the human body, or, as one academic put it to me, a club sandwich.

But, very quickly, the subject took on an added urgency, as I watched our assault on nature (by which I mean wildlife, plants, trees, habitats), as well as anthropogenic global warming leading to climate instability, heat up in real time. The question that drove me quickly flipped around: In what way does our disconnection from the natural world affect our mental health, our minds, our emotional lives? And how will climate chaos, extinction and environmental degradation affect the human spirit? This took me to the Svalbard Global Seed Vault in Spitsbergen in the Arctic, a man-made deep-freeze containing around 900,000 seed samples to protect agricultural biodiversity in the face of climate breakdown, nuclear war and natural disasters, and to Białowieża in Poland, one of Europe's last remaining primeval forests, which has been subject to intensive logging.

My travels, reading and interviews took in a broad gamut of ideas, desire lines and thinkers, from walking the South Downs with the Chief of the Order of Bards, Ovates and Druids to learning from ecotherapists practising in Canada, London and a secure-unit hospital in England, and to analysing the writing of, among others, Derek Jarman, Vievee Francis, Carl Jung and Edward Thomas to understand how nature heals.

The more I read and walked and heard, the more it dawned on me: if we lose our relationship with the natural world, we may, in some way, be losing a part of ourselves and a profound psychic experience that we all need. Do humans need the tonic of the wild to be happy? Do we require nature's goods and services not just to stay alive physically, but on a mental, emotional and spiritual level as well? Does a reduction in the time many of us now spend outdoors lead to psychic ills? In the twentieth century, we were ignorant about the potential of, for example, harm to the environment from chemicals and pesticides, and, more recently, plastics. Is our increasing separation from nature bad for our minds, too?

# 1

# Old Friends

Orchid—breathing
incense into
butterfly's wings
—MATSUO BASHŌ

I WAS THIRTY-TWO when I first used a rake, and dug my hands deep into soil. In 2016, three weeks after my daughter was born, we moved out of London to a town in the British countryside and, for the first time in my adult life, I lived in a house with a garden. Soon after, by some luck, a nearby allotment became free.

Starting off easy, we planted some manky old potatoes from our vegetable bag that had started sprouting green nodules. A few months later, I wrenched up the plants that had grown half a metre high and started digging. Avenues of black soil revealed moon after moon of luminous potato. So I planted more: an abundance of acid-green parsley, a blush of bright pink radishes, trumpets of butternut squash flowers spilling over and tumbling here and there.

As my time outside increased, I quickly noticed two things. First, my baby daughter seemed to like eating soil. Second, during and after time spent in the allotment or the garden, I felt happy, upbeat, less stressed and generally more positive. Initially, I put it down to the physical exercise, time to myself, the curious magic of botany, another date venue for my rekindled relationship with the natural world. In fact, there was likely to be a biological reason too, at least in part. I saw a poster on a Facebook parenting group. "Get Dirty," it proclaimed. "Exposure to soil bacteria

*Mycobacterium vaccae* is like a natural antidepressant, activating brain cells that improve mood, reduce anxiety and facilitate learning." "True or woo?" a user had asked the group. Most responded with anecdotal stories, and though there was a good deal of scepticism, someone wrote that, as far back as the 1760s, soil was thought to have a curative effect on the mentally ill.

In 2004, Mary O'Brien, an oncologist at the Royal Marsden in Surrey, England, discovered something fascinating by accident. She created a serum that contained *M. vaccae,* a species of bacteria found in soil. In close-up photographs, colonies look like mouldy, spotty, yellow growths. She wanted to see if the bacterium could boost the immune systems of her lung cancer patients and thus prolong their lives, because of its immunoregulatory effects, which were discovered in the 1990s. It didn't help them to live longer, but, strangely, those who received the immunization reported feeling happier.

Separately, a neuroscientist called Dr. Christopher Lowry was working at the University of Bristol, studying the antidepressant-like effects of the bacterium *M. vaccae*. He heard about O'Brien's findings and began to hypothesize that an immune response to *M. vaccae* stimulates the brain to create more serotonin, the happy chemical that antidepressant pills are designed to boost. He immunized mice with *M. vaccae* to find out more and reported that the immunized mice had a response to the bacterium, which could communicate to the brain and activate a group of serotonin neurons in the dorsal raphe nucleus, a structure in the midline of the brainstem. Inside this nucleus, serotonin-releasing cells are linked directly to the limbic system, where emotions are generated. This system is thought to play a crucial role in coping with stress. He tested the mice's stress levels by dropping them in a little swimming pool. Happy mice swim; stressed mice don't, according to previous research—and these *M. vaccae* mice enjoyed a dip. At the time Lowry told the BBC, "These studies help us understand how the body communicates with the brain and why a healthy immune system is important for maintaining mental health . . . They also leave us wondering if we shouldn't all spend more time playing in the dirt."

Since then, Lowry has spent years studying the impact of *M. vaccae* as a clinical application. Shifting his focus from allergies to psychological disorders, Lowry and his team wondered if *M. vaccae* could suppress inappropriate inflammation within cells, thus preventing negative outcomes of stress and post-traumatic stress disorder (PTSD)–like syndromes.

And it did. Mice injected with the bacterium exhibited fewer anxiety- or fear-like behaviours and were 50 percent less likely to have stress-induced colitis. Currently Lowry is trialling the effects of immuno-regulatory bacteria on people with PTSD, to see if it could buffer the side effects from high-stress situations such as combat.

The sixty-four-thousand-dollar question, as Lowry put it when we spoke, was exactly how bacteria like *M. vaccae* impact the brain to increase stress resilience. Scientists were studying sensory pathways to find out more. One possibility is that *M. vaccae* changes the phenotype (the physical properties and characteristics) of immune cells which migrate to the brain and regulate emotional behaviour. Lowry and his team also identified a small molecule in the bacterium which could prevent allergic asthma when injected. "We suspect that this is just one molecule out of hundreds," he said. "If you think about the whole scale of our microbiome it's unimaginable how many molecules like this there must be."

INSIDE YOUR BODY, there are probably more microbial cells than human cells. Symbiotic organisms colonize various areas of the body—the mouth, skin, vagina, pancreas, eyes and lungs—and many reside in the gut microbiota. You almost certainly have microscopic mites living on your face in the hundreds, or even thousands—mating, laying eggs and, at the end of their lives, exploding, unbeknown to you.

You may have heard the incredible fact that the resident microbes in your body outnumber your own human cells ten to one. That figure has been downgraded to three to one or an equal number, which is still astonishing. They mostly resemble mini jumping beans or Tic Tacs on a much smaller scale. These organisms aren't simply parasitic freeloaders: they are intricate networks that intertwine and interconnect, influencing our health and well-being through complex ecological processes. They are involved in the workings of the immune system, the gut-brain axis, protection against harmful organisms and, indirectly, they have some relationship to our mental health.

When we breathe, we suck different species of microorganisms into the body. Studies suggest fifty different species of mycobacteria would be normal in the upper airways of healthy individuals, making their way into the teeth, oral cavity and pharynx. The environment around you might look clear and empty, but it will be swarming with microscopic organisms, depending on where you are.

Our microbiota are healthiest when they are diverse—and a diverse microbiota is influenced positively by an environment filled with organisms, which are found more abundantly in outside spaces than inside. We imagine our skin and our bodies to be armoured, or a shell impenetrable to the outdoors, that we have somehow transcended our biological origins. But the human epidermis is more like a pond surface or a forest soil, as Paul Shepard, the late American environmentalist, suggested. Even if we don't yet understand or know exactly how many of the abundant microorganisms in our bodies arrived with us through exposure to nature— and, indeed, how they affect our mental and physical health—we are woven into the land, and wider ecosystems, more than we realize.

Crucially, these "old friends" that we have evolved with are able to treat or block chronic inflammation. There are two types of inflammation: the good, normal, protective type, whereby the immune system fires up to respond to an injury, with fever or swelling or redness; then there is the chronic, systemic kind you don't want. This is the simmering, low-level constant inflammation within the body which can lead to cardiovascular disease, inflammatory disorders, decreased resistance to stress and depression. This kind of raised, background inflammation is common in people who live in industrialized, urban environments and is associated with the unhealthy habits of the modern world: our diets, poor sleep, smoking and alcohol consumption, stress and sedentary lifestyles. As we age, our bodies become more inflamed. Scientists can measure levels of inflammation by looking at biomarkers such as proteins in the blood.

It should be no surprise, then, to learn that the gut microbiota of people who live in urban areas and developed countries are less biodiverse than those who still have profound contact with the land, such as hunter-gatherers and traditional farming communities.

Scientists are starting to understand more deeply the role inflammation may also play in our mental health. Evidence that bodily inflammation can affect the brain and have a direct effect on mood, cognition and behaviour is relatively new. But it is strong and compelling. Depression may well be all in the mind, the brain *and* the body. This view runs counter to the dominant view of Western medicine that our bodies and minds are separate and thus should be treated apart from each other, a view dating back to seventeenth-century French philosopher René Descartes' concept of dualism. As the neuropsychiatrist professor Edward Bullmore has said, "In Britain in 2018, the NHS is still planned on Cartesian lines. Patients literally go through different doors, attend different hospitals, to consult differently trained doctors, about their dualistically divided bodies."

But perhaps we are not as dualistically divided as the Cartesian orthodoxy our health systems are still built on would lead us to believe. A study of fifteen thousand children in England found that those who were inflamed at the age of nine were more likely to be depressed a decade later, as eighteen-year-olds. People with depression, anxiety, schizophrenia and other neuropsychiatric disorders have been found to have higher levels of inflammation biomarkers. European people have higher levels of cytokines in the winter months, which is also a time of increased risk of depression. Levels of cytokines are higher in sufferers of bipolar disorders during their manic episodes, and lower when they're in remission. Early findings suggest anti-inflammatory medicines may improve depressive symptoms. People with a dysregulated immune system are more likely to have psychiatric disorders.

In his book *The Inflamed Mind,* Bullmore argued that some depressions may be a symptom of inflammatory disease, directly related to high levels of cytokines in the blood, or a "cytokine squall," as he puts it.

Could our lack of contact with the natural world be a contributing factor to high levels of inflammation, which could be related to depression and other mental health disorders? Studies show that just two hours in a forest can significantly lower cytokine levels in the blood, soothing inflammation. This could partly be caused by exposure to important microorganisms.

There are multiple reasons why babies born in the rich, developed world have a less diverse population of mycobacteria—for example, the use of antibiotics, diet, lack of breastfeeding and reduced contact with the natural environment. We live inside, often in air-conditioned buildings cleaned with antibacterial sprays, with reduced exposure to organisms from the natural environment via plants, animals and the soil. Our food is sprayed and wrapped in plastic. We don't live alongside other species of animals, as we did for millennia. The opportunities to be exposed to diverse microorganisms are much fewer—which might explain why my daughter liked to eat soil.

"All babies left to their own devices will eat soil," said Graham Rook, Emeritus Professor of Medical Microbiology at University College London, and a distinguished expert on microbes and the immune system. "It's something that all vertebrates do." Rook has spent much of his long career studying how much the beneficial effect of green spaces on long-term human health outcomes might have to do with microbial biodiversity driving immunoregulatory mechanisms. We spoke over Skype and he was animated, owlish and quick to laugh. I enjoyed talking to him a lot.

Currently, scientists don't know exactly how many of the microbes in the soil also live within us, or which strains. Spores—reproductive cells that can develop into new bacteria—are also important. If you've got a low level of diverse gut microbiota, you can replenish them if they're encountered as spores in your environment, Rook told me. A third of all organisms in the gut are spore-forming. They're small, tough and can stick around in the environment for tens of thousands, maybe even millions of years. "Julius Caesar passed through this area," said Rook, speaking from his study in Islington, North London, "and perhaps there are spores from an organism from Julius Caesar's gut wafting around. Some of them might have colonized my gut. The imperial organisms," he joked.

Studies show that contact with natural environments during pregnancy or the neonatal period results in a lower prevalence of allergic disorder, which is connected with regulation of the immune system. Most of these studies have compared people who live on farms with those who don't. "Not only are they less likely to get allergic disorders, they're less likely to get some of the other chronic inflammatory disorders and probably less likely to have psychiatric problems as well," said Rook.

I don't think it was a coincidence that during the writing of this chapter I acquired a couple of backyard chickens.

THE AMISH AND THE HUTTERITES ARE traditional farming communities in the United States and Canada. The Amish population emigrated from Switzerland and the Hutterites from South Tyrol in Italy during the eighteenth and nineteenth centuries.

The Hutterites' belief in absolute pacifism and refusal to take part in military activities or pay war taxes has led to their persecution and banishment from various countries since their formation as a religious group in the sixteenth century. In the 1870s, the remaining population of around four hundred found a new home in Dakota County, Minnesota. In contrast to the Amish, the Hutterites live on large, industrialized farms. They don't shun technology, and farm with huge air-conditioned tractors. Because the barns are full of machinery, the children don't run around inside them.

The Amish, however, still live on traditional farms, using horses and ploughs, and the children play in the barns, spending time with the livestock, exactly as they did in the nineteenth century. They live across the United States but with large populations concentrated in Pennsylvania and Ohio.

A study of the innate immunity of children in the Amish and Hutterite communities found that the Amish had lower levels of chronic inflammation (the type you don't want, the subtle, long-term activation of the immune system). Just the dust around the Amish homes and farmyard areas was enough to protect against allergic asthma and other allergies. The prevalence of asthma in Amish versus Hutterite schoolchildren was 5.2 percent versus 21.3 percent and the prevalence of allergic sensitization was 7.2 percent versus 33.3 percent. It was a significant difference. The study also found that cytokines were much lower in the Amish, who were exposed to natural environments more than the Hutterites.

The study validates Lowry's earlier statement about us all needing to spend more time in the dirt. Perhaps, instead of teaching children that soil and the bugs that live in them are dirty and gross, we should encourage them, from a young age, to explore and play and get muddy, to enable them to benefit from the complex ecological networks—and rewards in terms of health—that exist within them. "Relating to depression, all we can say is that if you've got persistently raised background inflammation, you're more susceptible to it," said Rook. "So, for instance, I wouldn't mind betting that the Hutterites have more trouble with depression than the Amish. A psychologist would say, 'Oh, it's because of the wonderful relaxing effect of sitting on a plough watching a horse's arse in front of you for hours a day.' And they might be right, there may be some element of that, but I think that there are also these molecular mechanisms at work."

In affluent, developed countries, rates of allergies, autoimmune disorders and inflammatory bowel disorders have increased, suggesting that our immune systems are attacking targets they should not attack. Also, there are permanently raised levels of inflammation—seen in higher levels of the molecule C-reactive protein, a ring-shaped protein found in plasma which rises in response to inflammation—without evidence of inflammatory disorder.

What does it mean, then, to live in a city, as the vast majority of us do? By 2050, 68 percent of the world's population will live in urban areas.

In 2018, a study group that included Lowry and Rook compared people who grew up in the countryside around farm animals with those who grew up in cities without pets. The hypothesis was that growing up in a rural environment with greater microbial diversity would decrease vulnerability to stress-associated mental and physical health problems in later life.

The scientists gave twenty healthy young men from the two very different environments the Trier Social Stress Test to activate a stress and anxiety response. The procedure is used to induce a psycho-physiological stress response in a laboratory setting. Participants were asked to prepare a speech for a simulated job interview. In the test room, they were given no instructions about how long it should be. Then there was a mental arithmetic test, where the participant was asked to count backwards from 3,079 by subtraction of 17. If they made a mistake, they'd have to start from the beginning.

In the results of the test, those who grew up in the city had a higher social stress response, which was seen in exaggerated numbers of white blood cells and pro-inflammatory cytokines in the blood. This greater number suggests an increased risk for chronic, low-grade inflammation. "These data are in line with the biodiversity, missing-microbes, and old-friends hypotheses, which propose that the rapid rise in inflammatory physical and mental diseases in modern societies is due in part to a lack of exposure to immunoregulatory microorganisms," the study group concluded.

It made playing in the dirt sound even more critical, and I asked Lowry how confident he was in saying that spending time in nature, in the garden, with soil, could be both a treatment and a preventive measure for people with mental health problems. "I think that's very likely to be true," he said. "Then the question is, what's the effect size? How big is the effect? It would be hard to argue that that kind of exposure would not be beneficial, but how beneficial compared with other interventions? I think we might be surprised how effective that might be."

And it is not a quick, temporary fix. In animal studies, T-cells (cells which modulate the immune system) have a surprisingly long half-life of twenty-seven days. This suggests that your post-gardening high could keep you buzzing for days, or even weeks.

AFTER READING ABOUT *M. vaccae* I had a strong urge to touch and smell soil. I biked to a nearby area of woodland on the Galloway coastline in south-west Scotland, where I was staying at my mother's. I parked up on one of the trail paths, bum-shuffled down the bank and entered the forest, avoiding boggy patches and fallen logs. It was a late autumn afternoon and the sun was inning and outing, larking through the trunks. Bark flaked like scabs and the trees had a dusting of moss. The birds' vespers were soft and lazy and I could hear running water. The carpet of the forest was the thing that made it so special—effulgent Kermit green with bright moss, tiny velvet matte ferns and three-leaf clovers. I plonked myself down, tore apart a handful of ground and *snifffed*. The soil smelled like bacon, whisky, dill, gravad lax, gherkin and peat. It crumbled like cake in my fingers.

Even if you haven't sniffed a handful of soil up close, you will most likely have smelled petrichor, the delicious earthy scent after rainfall, when oil from plants is released into the air. In 2007, chemists at Brown University in the United States discovered the organic compound responsible for the metallic smell of soil or earth, which is more pronounced after it has rained. It is called geosmin.

Humans are acutely sensitive to the smell and can detect low concentrations of geosmin at five parts per trillion. There may be an evolutionary explanation for our sensitivity. Some suggest that the reason we're so attuned to the earth's perfume is because our hunter-gatherer ancestors would have followed their noses to find rainy, irrigated landscapes for food and survival.

Early studies on how geosmin may affect the brain are starting to be published. A paper by a team at the School of Natural Resources and Environmental Sciences in South Korea looked at the effect of smelling geosmin on brain activity of men and women. The study didn't have a placebo control and it was a small sample size, but the team concluded that the brainwave activity was linked with calmness and relaxation states of the brain, particularly in women.

I returned to the same woodland the next day, wondering if I might camp there one night. I slipped away from the track, past the gnarled fallen tree and into the Emerald City. It had rained that day but the sun triumphed through the charcoal skies. I stood in the calm, quiet lull of the forest, listening to the bristle of the branches and the occasional bird call, smelling the aroma of the earth and marvelling at raindrops like diamonds on the moss. To my left, I sensed movement. I turned slightly, and found I was face to face with a female roe deer. She paused, as did I, and we surveyed each other. I was so close I could see the wet of her nose and strands of soft tufts in her ear. My heart beat faster, as I imagine hers did too, and I could feel adrenaline bolting through my limbs, perhaps a physiological remnant from our hunter-gatherer times. She dipped her pretty eyes and then her neck and leapt off, softly padding the forest floor, vanishing between the trees.

PART II

# ROOTS

∽

2

# Biophilia

What if, long after all of nature has finally been ground up in the garbage disposal of the technologic sink, it becomes clear that there are indispensable genetic needs for many of these components of nature?
                    —HUGH ILTIS, Professor of Botany, in a letter to *Science,* 1973

Thousands of tired, nerve-shaken, over-civilised people are beginning to find out that going to the mountains is going home; wildness is a necessity.                    —JOHN MUIR, *Our National Parks,* 1901

EDWARD OSBORNE WILSON WAS fishing off Paradise Beach in the Panhandle of northern Florida as a young teenager in the early 1940s when a silvery perch caught his eye. Speckled and spawning, the fish glinted in the sunlight. Wilson struck with accuracy and brawn: the perch flew out of the water and its spines shot straight into his right eye. The surgeons were unable to save his eye and he was left with just close-range vision in the other. The keen young naturalist had to abandon ambitions of studying birds, frogs or bears and look for something nearer and smaller. First, he decided to investigate flies, but a shortage of pins after the Second World War interrupted his work. His next idea was ants. In time, Wilson became obsessed with the social lives of ants and went on to publish many award-winning works and biological studies. He won the Pulitzer Prize twice for his books *On Human Nature* (1979) and *The Ants* (with Bert Hölldobler, 1990) and the National Medal of Science in 1976. His 1984 book *Biophilia* catalysed the ecotherapy movement, biophilic architecture and design and the wider scientific field of nature and health. The book itself is quite obscure—you won't find it in many high-street bookshops today—but the word and concept appeared repeatedly at nature and health conferences I attended, in studies, papers and journals I read, and in my interviews with both scientists and non-scientists in the field. I ordered it from the British Library and travelled to London to read it.

The text is a mixture of things: a call for a new conservation ethic in the unfurling biodiversity crisis, noticeable even in 1984; a meditation on science and art; and a travelogue of Wilson's journeys across the world. At its heart, it poses the question: Is human sanity at risk in a world of depleted nature?

The biophilia concept is Wilson's notion that humans have an innate and emotional affiliation to life, life-like processes and other living organisms. Our tendency to be drawn towards living things—from asking our parents for a pet hamster to leaving flowers on the grave of a loved one—is, according to Wilson, the expression of a biological need with a genetic basis. Humans depend on nature for more than food, he argues. We have an evolutionary need to connect with the natural world for cognitive, mental, emotional and spiritual development, growth, meaning and fulfilment. Without contact with the natural world, we become impoverished.


WILSON DOESN'T PRESENT evidence in a formal empirical sense in *Biophilia*. It was—and remains—a hypothesis. He argues that the biophilic tendency was "so clearly evinced in daily life and widely distributed" that it deserved serious attention. It couldn't be a coincidence that humans have always loved, worshipped and celebrated nature and animals. "The biophilic tendency . . . cascades into repetitive patterns of culture across most or all societies . . . they are too consistent to be dismissed as the result of purely historical events working on a mental blank slate."

Where we evolved is Wilson's first piece of the argument. He begins from a position of evolutionary logic, looking at the original human environment, where our brains evolved over about two million years, during which people lived in intimate contact with the natural world. "Snakes mattered. The smell of water, the hum of a bee, the directional bend of a plant stalk mattered." He posits that because 99 percent of human history was spent in hunter-gatherer groups that were intimately related to other organisms, predators, prey and habitats—a biocentric world, rather than a machine-regulated world—it is likely that genes prescribing the learning propensities that improve chances of survival, food or reproductive fitness were spread by natural selection. "A certain genotype makes a behavioural response more likely, the response enhances survival and reproductive fitness, the genotype consequently spreads through the population, and the behavioural response grows more frequent," he explained.

In other words, "we learn what we know, but some things are learned more quickly and easily than others." And those things are likely to be related to the natural environment in which *Homo sapiens* lived for the great majority of its history.

The second piece of evidence follows on from this. Choosing the right place to live, where there would be ample food for the tribal group, as well as water and shelter, was crucial to the survival of our species for millennia. It hasn't been vital only for the tiniest sliver of our evolutionary history. If our primary function was to select a good place to live, suggests Wilson, it is probable that our brains and senses will have evolved characteristics that would help with that. The modern human—you, me, us—doesn't arrive on Earth as if stepping off a train. Our flesh and DNA and thoughts and preferences are marked by the past.

Wilson points out that when people have the choice today, they will choose to live in, or spend time in, landscapes with the same key natural characteristics: large, park-like grasslands, with clusters of groves and trees, and water. These were the type of environs where ancient humans lived and were most likely to survive, and where our senses, intellect, emotions and cognition evolved. Humans, with their bipedal locomotion and free-swinging arms, were suited to the open plains, the savannahs, first of Africa and then of Europe and Asia, where they could hunt and chase game, and pick, dig and pluck fruits and tubers. We didn't settle in rainforests, or deserts, and the "worldwide tendency" to gravitate towards savannah-like landscapes which resemble the habitats of our ancestors, or, for those who can afford it, homes, temples and palaces that look out over lakes, rivers or the sea, is evidence to Wilson of a gene that selects for those kind of habitats. The landscape architects and gardeners who base their designs on this savannah aesthetic are responding to a "deep genetic memory of mankind's optimal environment."

This idea was tested by Gordon Orians, a writer and Professor Emeritus of Biology at the University of Washington, by looking through the "before" and "after" plans and drawings of the eighteenth-century British landscape architect Sir Humphry Repton, who designed the grounds and gardens of private estates and country houses. Working with the American psychologist Judith H. Heerwagen, Orians hypothesized that the "after" plans would be more "savannah-like" with scattered trees or copses in open vistas. They would also include refuges (huts, houses), water and animals. The researchers found that Repton had added small clusters of trees to almost half of his designs, often into open spaces. He also thinned and removed trees in nearly half of the landscapes, to give better views. Water features, boats and grazing animals were also added. "These features are all characteristic of savannah environments, the habitat in which humans lived for millions of years," they wrote.

Decades later, scientists are still using habitat theory, as it is called, to test the biophilia hypothesis, and still certain preferences come up. At the International Conference on Quality of Life at the Pacific Sutera Hotel in Sabah, Malaysia, in 2014, scientists from the Faculty of Built Environment and Surveying at Universiti Teknologi Malaysia presented the results of a study that supported the hypothesis.

The research team showed five photographs of areas in the Melaka reserved forest to fifty-one experts—students with a background in landscape architecture or urban planning—and 126 non-experts from the local population. The participants were then asked how they perceived the scene, using the Likert scale, which measures intensity of feeling. The factors they were asked to rate ranged from mystery to coherence, and stewardship to disturbance.

The study identified three dominant attributes that the group—both experts and non-experts, and ignoring demographic differences—used to assess the landscape: complexity (the diversity and richness of natural features), naturalness (the level of wild, untouched landscape) and legibility (ease of finding one's way, in a strange, unknown place, with a landmark or feature).

The team concluded that because these three aspects were strongly preferred across the groups, the results gave support to the notion that there is a common connection that bonds humans, stretching far back into our ancient past, but remaining in our DNA to this day.

. . .

IN THE EARLY WEEKS of motherhood, I found a beautiful, wild cemetery containing thirteenth-century church ruins next to our new home. At its centre, an enormous, majestic beech tree shone batter-yellow in the early autumn sun. In its shadow, under the neuron branches, was a pink lake of cyclamens. Rabbits flickered in the long grass, occasionally stopping long enough so that I could see the black inkwells of their eyes. Goblets of ancient epiphytes decorated the brick walls. When I looked closer at the yellow blotches of lichen with my pocket microscope, I could see tiny cities made of gold, with depth and dimension and cherry-red microscopic bugs invisible to the naked eye. Drug users would sometimes huddle behind tombstones and occasionally I saw needles stuck into trunks. In those early months, I was still in survival mode and could barely acknowledge anything that wasn't to do with keeping my daughter alive and well. But there was something about my daily walks that lifted me out of myself.

When I walked up the slight hill into the cemetery, underneath a large, low-slung, spreading canopy of branches, I felt calmer. It happened every time: as soon as I was under the dark green of the yew's cover, it was as if I'd taken a split-second yoga class. I would opt for that route in—there were three entrances, and another one was a much nicer walk via allotments rather than a busy road—to experience the momentary release of tension. I wondered if it could be a primal attraction to elements associated with shelter, as well as food and water—basic requirements for a new baby.

Perhaps some genetic predisposition made me gravitate towards this shadowy corridor—and it may have had something to do with the specific shape of the tree. For millennia, trees have played a crucial part in human survival and well-being. They have provided shelter—somewhere to sleep, or rest under—and food, materials and medicine. They also allowed hunter-gatherers a view over the surrounding area to seek water, food and further shelter, or to observe the presence of predators. Orians studied tree preferences as well as habitat selection and set out to answer a question: Are all trees equally effective at producing positive responses? He proposed that the most pleasing shapes would mirror the trees which

helped humans during our evolutionary history. On the African savannah, a characteristic tree is the *Acacia tortilis,* which has a broad canopy
and spreads in width further than the tree is tall, with layered, umbrella-
like branches and small leaves.

To test his hypothesis, Orians showed a group of people a large number of photographs of trees that varied in trunk length, canopy density,
canopy layering and broadness, compared with height. He found that,
to this day, people still prefer trees with highly layered canopies, lower
trunks and a higher canopy width–to–tree height ratio. These trees
are easier to hide under, or to shelter beneath from the sun, or to climb
up to safety—exactly like the tree at the entrance to my local cemetery
park. "The results from our preliminary survey are in accordance with
a functional-evolutionary perspective on the relationship between trees
and humans," concluded Orians and Heerwagen.

The time of day is another factor which early humans would have
been acutely aware of. Without light, they would have been unable to
see a wolf or bear ready to pounce, or a snake near the camp or cave. So
they would want to make sure they were firmly ensconced in a safe and
protected habitat as the sun began to set.

Orians and Heerwagen looked at landscape paintings of sunsets,
predicting in the light of their hypothesis that they would be "high in
refuge symbolism," featuring a house, a church or other dwelling. Their
predictions proved correct: 66 percent of the sunset paintings they surveyed featured a refuge, and in 92 percent of cases the refuge was highly
accessible.

The poet and geographer Jay Appleton built on this work and set out
his theory of prospect-refuge in his 1975 book *The Experience of Landscape.* He wrote that the ideal environment for early humans was one
where they could hide from predators (refuge) while being able to see
danger and potential prey (prospect). Both Appleton and Orians propose
that we modern humans have an "inborn desire" to be in these types of
landscape, a genetic predisposition to prefer landscapes and natural settings where we can see a horizon or survey from a vantage point across a
wide, open space and the surrounding landscape; where there are small
copses of trees, which offer refuge, food and shelter and, ideally, the presence of water.

In tribute to the recently deceased Appleton, Guy Lochhead, a writer for the journal *Ernest,* a magazine about "meandering journeys" and "wild ideas," looked up the top one hundred images of landscapes on the website Flickr to spot patterns in the places people prefer to photograph. He found that 91 percent of the images included some form of refuge; 99 percent contained some symbol of prospect; and 96 percent included both. Without our realizing it, our genes may well govern our aesthetic preferences.

THE NEXT PIECE OF EVIDENCE for E. O. Wilson's biophilia concept has to do with the animacy of life. From a baby's earliest weeks and months, when it can barely focus its eyes, or see colours that aren't black or white, it learns what is alive and what is inanimate. Wilson argues that humans react more quickly and fully to living organisms because life of any kind is more interesting than that which is inert and motionless. People "prefer entities that are complicated, growing, and sufficiently unpredictable to be interesting." A squirrel is more interesting to watch than a crisp packet, a butterfly more so than a traffic cone.

A study of newborn babies from 2008 supports Wilson's argument. It looked at preference for biological activity over non-biological activity. Babies just two days old were shown images of random dots and the movement of a walking hen. The babies preferred to watch the hen, or the "biological motion display," as the scientists poetically put it. They tested babies who had just been born to see if they had an intrinsic, innate perception of biological motion, and a preference for it, before prepared learning came in at a later stage of infancy. The results suggested that we may have evolved visual perception that allows us to attend particularly to other animals or the movements of another animal (surely the mother, as the source of food).

The fact that we are attracted to moving objects or landscapes that are pleasant and beautiful may seem obvious, but, as Wilson writes, the "obvious is usually profoundly significant." A landscape isn't just nice because it's nice, but because its meaning is "rooted in the distant genetic past." It may not be ground-breaking to say that most people would prefer to look at a tree than a pile of dead leaves, but then consider the boxes we trap ourselves in, and how rarely, now, we seek the abundance of natural diversity. If it is so obvious, why aren't we doing more to protect the living, breathing, running, squirming, jumping, dancing, spinning, glowing natural world?

AN ANIMAL MAY BE more stimulating than an inanimate object, a natural environment more pleasing than a built one—and it may well contain some helpful microbacteria that could improve our health—but does exposure to nature make a *measurable* difference to the way we feel or behave or recover?

One of the earliest scientific studies into nature and health took place at a suburban hospital in Pennsylvania. Roger Ulrich, Professor of Architecture at the Center for Healthcare Building Research at Chalmers University of Technology in Sweden, and now the leading expert on evidence-based healthcare design, looked into the records of forty-six patients recovering from gall bladder surgery between 1972 and 1981: he wanted to investigate whether a natural view out of a window had a more restorative effect than an urban view. Ulrich himself had suffered from kidney disease in his teenage years and considered his view of a pine tree while he was bedridden to have aided his recovery. The only difference between the two types of hospital rooms was the view out of the window, which the patients could see from their beds. Rooms on one side of the building looked out onto deciduous trees; the others looked out onto a brick wall. The patients were assigned to rooms as they became vacant. Crucially, they didn't have a choice about which type of view they could look onto, which removed the possibility that people who liked looking at greenery would request rooms with a view of the trees.

Ulrich looked at the hospital records of the patients who had had cholecystectomies between 1972 to 1981 in May and October. The five areas of data were the number of days in hospital, the number and strength of painkillers and anti-anxiety drugs given each day, minor post-surgery complications and the nurses' own notes on recovery. These were filed into two groups: positive ("in good spirits") or negative ("crying," "upset," "needs encouragement").

The results were remarkable. The patients with the views of trees had shorter post-operative hospital stays, fewer negative evaluative comments from nurses, and took fewer moderate and strong analgesic doses. They also had slightly lower scores for minor complications. There was no significant variation between the groups on anti-anxiety drugs, though Ulrich suggested that might be because the wall-view patients had more narcotic analgesics, which might have sedated them enough not to need tranquillizers.

In his notes, Ulrich conceded that as urban scenes go, a brick wall is the most monotonous, and not all built views may have the same results. One would imagine that a view of St. Mark's Square in Venice or of a gold-stone village in Provence or of the magnificent skyline of New York could be therapeutic for some—but the research findings were acted upon and influenced hospital design in the United States and Britain, where Ulrich served as an adviser for the NHS during the 2000s. It also influenced many others to look deeper into the associations between nature and health and build on Ulrich's work.

In the past twenty years, studies conducted in countries across the world found similar positive benefits of a connection to nature on mental health; for example, that the presence of trees and greenery translates to fewer antidepressant prescriptions. A study in London used two sets of data points: first, a census of street trees; and second, NHS data on antidepressant prescribing. The authors found a qualitative relationship between tree canopy and antidepressant prescription. Of course, it's not as simple as "plant a tree and ditch your pills," but these studies—the number of which has grown significantly throughout the early twenty-first century—have looked at everything from the effect of birdsong to gardening, woodlands to living by the coast, exposure to parks to outdoor learning, and their relationship to physical and mental health. They suggest that exposure to the more-than-human world has an important relationship to human mental health that has often been overlooked.

Thousands of scientists are now investigating the relationship between people and the natural environment—or lack of it—and, in doing so, the quality of evidence is improving to the point where policy-makers and healthcare professionals are slowly, slowly taking notice. From October 2018, doctors on the Shetland Islands began to offer "nature prescriptions" to patients with mental illness, diabetes, stress and heart diseases as a supplement to other treatments. The prescriptions take the form of calendars and walking routes devised by the Royal Society for the Protection of Birds (RSPB) where lapwings, fulmars and oystercatchers can be seen. In the United States, between April 2018 and early 2019, 169 health professionals were issuing "park" prescriptions.

The highest standard of evidence is required for a treatment to be adopted by a healthcare system. And if a council is deciding whether to allocate money to maintain the tree canopy in a public walkway in preference to fire services or crime reduction, it will need the proof that a level of greenery will translate to better health for its citizens.

But isn't it strange that we have become so detached from the natural world that we need that proof laid out for us? That we hold, for example, nature-based interventions to the same standard as pharmacology and thus only a handful of GPs in Britain will dare to prescribe it? That we don't consider our relationship with the environment as a significant part of the larger biopsychosocial picture of human health? And that, nowadays, we need to be told by a professional to go outside? Isn't it bizarre, really, that we consider spending money on conserving the last scraps of nature in our urban areas a luxury? Although thousands of scientists are working to explore and test the notion that we need nature for the health of our minds, it is still not obvious to us that a meaningful relationship with the natural world is an important facet of human health while the health of the planet is rapidly declining in our hands.

BUT WHAT ABOUT those of us who don't like spending time in nature at all? Those like Woody Allen who would say, "I love nature, I just don't want to get any of it on me." Many of us choose to make our homes in urban areas, living alongside lots of other people, with little or poor connection to natural spaces, let alone access to a wilderness.

In an essay for *The Biophilia Hypothesis* (1993), the Indian ecologist Madhav Gadgil wrote that, "If there is a learning rule that inclines humans to love natural diversity, very few among the urban middle classes of India actually come to do so." He wrote about watching a process of "desacralisation of nature" in India and a shift from worshipping natural elements to man-made icons, as market forces spread through the country.

I contacted Gadgil to see whether his perspective was still the same twenty-five years later. Farmers, herders, fishermen and forest-produce gatherers in India were attached to the natural world for the benefit both to their livelihoods and also their psychological well-being, he told me. But urbanites were increasingly alienated. Apart from a small minority who engaged in trekking or paragliding, most city-dwellers spent more time on their computers or watching television than engaging with the natural world.

But even if people don't care much for spending time in nature, do they still need it, even if it's there in the background? Or, to put it another way, if people aren't aware of the grandeur of the original biosphere, is the loss of it a threat to their sanity? As many of us, increasingly, don't experience nature at all and choose to do other things, does it matter?

In 2009, an evaluation by biologists at the Norwegian Institute of Public Health of fifty empirical studies which had appeared in established scientific journals was published in the *International Journal of Environmental Research and Public Health*. Does biophilia have merit as a hypothesis, it asked, and does depletion of natural elements have a negative impact on the human mind? Unlike the move from cave floor to mattress—clearly a good thing—could the move from outside to inside involve unwanted negative consequences?

The paper detailed the field of exploration into how nature is linked with health and well-being, from reducing stress and mental restoration to fewer sick days and reduced attention deficit; the effects of gardens on health outcomes to responses to window views in the workplace (i.e., fewer sick days; less tiredness and coughing). In conclusion, the research team felt that the hypothesis had merit. First, the idea that interacting with the natural world could have a positive effect on the mind was "reasonably well-substantiated"; and second, it was "likely" that even if people had no connection with the natural world, out of choice and preference, absence of a connection to nature was harmful: "The biophilia trait can be reinforced or subdued by individual learning." And this was the part that worried me: "It seems likely, however, that even in individuals who do not express any appreciation for plants and nature, the lack of nature can have a negative effect."

If you look around, or read the papers or watch the news, you would imagine that biophilia in twenty-first-century humans must be weak, for we rampage over the rest of nature more than we care for or nurture it. Our treatment of the natural world would suggest an abusive relationship rather than one of love, care and reciprocity. Indeed, look at the way Prince Charles was mocked mercilessly for saying in an interview in 1986 that he talks to his plants. He has never lived it down, branded a "loony" and "potty" for revealing his loving relationship with nature. As the study of plant communication has advanced, and his forward-thinking action on organic farming and environmental issues has proved to be on the right side of history, Prince Charles's approach doesn't look so eccentric after all. In a later interview about the ridicule, he said that speaking to his plants keeps him "relatively sane." (And in a world with more understanding and sympathy for mental health problems, we probably wouldn't chortle at that now.)

Nowadays, it is not so clear that "people react more quickly and fully to organisms than to machines," as Wilson says. Our reliance on technology may have led to an attenuation of the human urge to connect with nature. Certainly, there are fewer opportunities for the biophilia gene to be triggered in modern-day Western society. We teach our children to be scared of spiders and repulsed by dirt. We pave our gardens and spray our flowers with pesticides. We don't even know the names of the creatures we live beside. It is not, for most, much of a relationship.

Responses to the natural world from two Uto-Aztecan cultures along the US-Mexico desert borderlands—the O'odham and the Yaqui—led anthropologists Gary Paul Nabhan and Sara St. Antoine to believe that the biophilia gene needed to be triggered. The children they surveyed, despite access to large, open, natural spaces, were more interested in watching television and had a much more tenuous relationship with the natural world than their parents and grandparents did. It was the "extinction of experience" in action, they reported back in 1993.

If biophilia is subdued, does it even matter? Can you miss what you never knew? Wilson feels strongly about this: "On Earth no less than in space, lawn grass, potted plants, caged parakeets, puppies, and rubber snakes are not enough," he writes, and compares nature-less people to monkeys in lab cages and cattle in feeding barns.

And if it isn't triggered, what are the long-term consequences? Is it possible that biophilic tendencies could die out, and fade from our genotype? Would they remain dormant if they were never activated? Could we be sleepwalking into a time when the natural world is reduced to its bare minimum (a time which wouldn't last very long)? If children aren't taken outside, and rarely have an opportunity to play in the woods, won't they just satisfy their need for restoration and stress-busting elsewhere, in bowling and ice cream, Netflix and shopping, and never know the difference? Will plastic trees and simulated virtual reality gardens be enough for future humans? Are we so desensitized that we are losing the thirst for a relationship with the natural world? And is the absence of this connection causing us harm, whether consciously or not? The juggernaut of therapeutic power I experienced from spending time in the natural world during my recovery from addiction—and the many similar stories I have read and heard—suggests to me that the answer is yes.

Over the phone, Wilson, then eighty-eight, told me that biophilia was unlikely to be bred out of us. "It takes a long time in populations for one set of genes to be replaced by another, to eliminate genes pretty well probably scattered through the genome that work together in creating mood and modes of response—what we call prepared learning, learning one thing and not another—we love flowers, we're easily scared by snakes."

Given that our relation to nature is as much a part of our history as social behaviour itself, it would be extraordinary to find that our learning of all the rules related to the world we evolved in has been erased in a few hundred years. "It would take some ugly eugenics to wipe out any regard and love of nature in a few generations. It takes hundreds of generations. We're not insane, we're not going to go that way; we're going to go the other way, and bring it back." Wilson was essentially optimistic, although he said there was still a way to go before we fully grasp the urgency. (In *Half-Earth,* published in 2016, he argued that the only solution to impending ecological collapse was to devote half the planet to the rest of nature.)

If we continue to take the path of detachment and divorce from nature, at a time of growing biodiversity loss and catastrophic global warming, the consequences of our eco-illiteracy will grow even more hazardous. Alongside the lack of relationship comes a lack of knowledge and fundamental ignorance about the ecosystems that keep the planet thriving. "Nature" isn't just beautiful and intriguing and awesome; it is our life-support system. If future generations don't care or aren't interested in the rest of nature, if they aren't taught about ecology and that we only eat and breathe because of plants, why would they bother with conservation or connection? If they haven't felt the soothing calm of a walk in the woods or heard the song of a nightingale, what will they miss, through ignorance? You can't love someone you don't know, or notice, or spend time with.

IT IS A GREY, cold January day back in my garden, but there is a pearly strip of spilt pink milk on the horizon. I am outside to clip the withered clematis which is obscuring the very little light available in our living room. It is harder to cut than I anticipated: even though it is the depths of winter, tiny tendrils have wrapped themselves around each other, around and around and around. They are adamantine and wiry and testament to the plant's desire to grow and move. At a couple of points, it has tried to burrow in through the windows, lodging itself into the plastic frames, ambitious even in these dormant days, driven to grow and flourish and restore. It makes me think of my very young daughter's need and drive to be outside, an innate desire to be in the fresh air, a desire that hasn't yet been distracted or influenced or cultured or educated out of her.

# 3

# Mud-Luscious and Puddle-Wonderful

> But best of all was the warm thick slobber
> Of frogspawn that grew like clotted water
> In the shade of the banks.
> —SEAMUS HEANEY, "Death of a Naturalist"

> In North America by the time you're eighteen years old, you've spent
> 12,000 hours in the classroom. That's 12,000 hours in a rectangular
> room separated from nature and that's kind of what it takes, I think, to
> create an industrial society.
> —JOHN SCULL, ecopsychologist, on Vancouver Island

> It was beneath the trees of the grounds belonging to our house, or on
> the bleak sides of the woodless mountains near, that my true composi-
> tions, the airy flights of my imagination, were born and fostered.
> —MARY SHELLEY, Introduction to *Frankenstein*

SUNLIGHT DANCED on the floor of a small wood near my grandparents' house in south-west Scotland. It was a warm day, unusually
warm for early April, and I tied my jumper around my waist as I
strolled over to the tree house, wiping cereal milk from my mouth, skip-
ping over in my muddy trainers, hoping I could get there before my little
brother started following me. I wanted to be free to ascend the sycamore
and sit in the tree house alone. I started to climb. I was in the canopy of
a rainforest, monkeys on my left, toucans on my right. I *was* a monkey,
in fact. Or I was Maid Marian, hiding from an army, bow and arrow on
my back, safe in a tree. Then, I was a leaf—safe for now, but waiting for a
puff of breeze to take me down. The tree creaked and swayed as I pulled

myself up. I looked out from my crow's nest, through the dense foliage of the Amazon jungle, my kingdom, a battle below, to Cair Paravel; across swamps and bogs to the low tide scattered with black blistered seaweed and tiny worm hills and scrunches of crabs in icy rock pools brightened by raspberry jelly anemones. To the left, dinosaurs roamed and then a tidal wave and a dragon emerged from the sea. But I was safe in the tree house. Safe in the tree. I remembered the task in hand. To make a secret magic potion out of pine cones and petals and shells and shampoo, mixed in a tiny cauldron with a great big stick. Then it was lunchtime—Granny's broth—and a nap with dreams of whales and worms and flying above the canopy on a giant lunar moth.

I wasn't there often, but for the couple of times a year I visited, the woods were my venue, my stage and the blank slate which nourished my imagination. A deep part of me was nurtured there. My cousins and I learned to be brave, to push ourselves and take risks, discovering that it was OK to get strung up by the ankles in a rope-swing even if someone didn't find you for what felt like a long day.

Home was a boarding school my father taught at, which was a reasonably safe place for a group of teachers' kids to roam around in, scampering over obstacle courses, up climbing frames and playing pooh sticks at a nearby stream. The grounds were mostly comprised of playing fields and an enchanting flowered garden with a hut we believed was the home of a troll. At the age of nine I went to a nearby primary school which had in its play area a number of small, sheltered wooded areas. My friends and I would run to our patch under the trees at break-time, looking for insects and secret treasure and larvae. We would make our area—"Oak Army Barracks"—tidy, sweeping the floor, moving the wooden log around, keeping our imaginary horses fed and watered, and changing their colours as we pleased.

My clearest and happiest childhood memories are of being outside. Lying under a big tree and listening to the teacher read aloud from a book, watching the clouds, feeling safe. The feel of wood chips. The witchetty grubs, all white and fat, that we found in a log and dyed pink with food colouring. Rabbits nibbling the grass on the rounders' field as the sun set. Stick insects. Frog spawn. Swinging from a tree and feeling free. Cold winter netball skies. Grass stains on my knees. My glamorous great-uncle from San Francisco teaching me how to extract nectar out of honeysuckle. Catching pond skaters. Conkers in their green-spiked wombs. Collecting bright green aphids and observing the gentle pulse of their wings. Collecting ladybirds. Collecting snails. My father stopping the car at night to listen to crickets and frogs on holiday in France. These are my favourite, lasting memories of being a child in the 1990s.

A couple of decades later and the landscape for children has changed dramatically. Three-quarters of children (aged five to twelve) in the United Kingdom now spend less time outdoors than prison inmates, who require, according to UN guidelines, at least one hour of exercise in the open air every day. We've moved far away from Rousseau's ideas of education, that "fresh air affects children's constitutions, particularly in early years. It enters every pore of a soft and tender skin, it has a powerful effect on their young bodies. Its effects can never be destroyed." Fewer than one in ten children regularly play in wild spaces now, and the area around a child's home where they can explore unsupervised has shrunk by nearly 90 percent since the 1970s. Break-time or recess has also decreased since the 1990s, in both the United Kingdom and the United States, which means less time to play and commune with the outdoor world. Children don't walk to school as much or as freely, although urbanites may be more likely to walk to class than rural children.

In 2013, the RSPB published a three-year study on children's connection to nature. It devised a sophisticated measuring system, based on four factors: enjoyment of nature; empathy for creatures; a sense of oneness with nature; and a sense of responsibility for the environment. The report concluded that four out of five children did not have an adequate connection with the natural world.

This is not only sad because their world won't be "mud-luscious" and "puddle-wonderful" when they're young—there are other ramifications. First, research shows that a connection with nature in childhood leads to a connection with nature in adulthood. If a child is introduced to the natural world before the age of twelve, the chances are they'll continue the relationship and its benefits into adulthood. It is crucial to get them early, as they say.

Secondly, without exposure as a child, particularly to wild nature, people are less likely to look after and protect natural areas as adults—leading to a potential catastrophe for nature conservation.

Although children spend much more time using screens than they do playing outside these days—one study found that children under seven spent twice as much time looking at screens (four hours a day) as they do outside (an hour and a half a day, which sounds like quite a lot to me)—technology itself isn't the sole reason behind this shift, although of course computers and a more sedentary lifestyle are contributing factors. Access to green space is restricted, and increasing numbers of roads, traffic and motorways mean it's not as safe to get around. A "safety first" culture driven by fear of injury, accident and violence has made parents less likely to allow their children out. As high-rise developments replaced terraced houses, street-play vanished. Children in low-income or ethnic minority communities have particularly limited access to natural spaces, with low-income communities receiving less funding for public parks. Teachers cite pressure on resources and exam results as reasons for not taking children outside for lessons.

In Britain, parks—the outdoor venue where most children play— are under threat. Bristol city council announced that spending on parks would be cut to zero in 2019. Newcastle council has cut its park budget by 90 percent over seven years. Other councils, such as Bexley, Waltham Forest and Stockport, are allowing development on urban green spaces. The State of UK Public Parks 2016 report found that 92 percent of local authority parks departments had experienced budget reductions, and 95 percent of parks managers expected to be faced with further reductions in the following three years. In 2015–16, local authorities spent just 0.73 percent of their budgets on parks.

It's almost as if we actively don't want children to play in natural outdoor environments. I don't think we believe, as the highly influential psychologist and American Psychological Association president Edward Thorndike argued in the 1910s, that teaching children to love nature is immoral. "Let us remember that it is not only not wrong for the child *not* to love the plants and flowers," he wrote. "It is really wrong for him to love them, for it is unreasonable and therefore mischievous idolatry." But one does wonder.


IN 2007, the words "acorn" and "buttercup" were taken out of the *Oxford Children's Dictionary,* in favour of words like "broadband" and "cut and paste" to reflect changing usage of the language. "Hamster," "heron," "herring," "kingfisher," "lark," "leopard," "lobster," "magpie," "minnow," "mussel," "newt," "otter," "ox," "oyster" and "panther" were also deemed archaic and removed. It took until 2014 for people to realize what had happened, and then a group of twenty-eight influential authors in the literary world, including Margaret Atwood, Andrew Motion and Sara Maitland, protested. At the expense of outdoor play, children were experiencing obesity, anti-social behaviour, friendlessness and fear, the authors wrote in their campaign letter.

The researcher and science writer Martin Robbins wrote a defence of the Oxford Dictionaries in *The Guardian* soon afterwards. He used Google's Ngram Viewer to look at how often words such as "fern," "catkin" and "buttercup" had appeared in literature over the last two centuries. The graphs showed decline. "It isn't the job of the OUP to get kids to play outside: that's called parenting, and maybe that's where campaigners should be focusing their attention," Robbins wrote. I think he was right to defend Oxford University Press. The fault doesn't lie with the dictionary's publisher, it lies with society at large: it presents a damning indictment of where we are now and what we prioritize, value and are distracted by.

The removal of the words from the dictionary simply reflected a further distancing from nature. Perhaps a more interesting question isn't whose fault it is, but whether the mental health of children is affected by a relationship with the rest of nature or, adversely, by a severance from it? As the writer Jay Griffiths has said, "Children . . . are denied the soul medicine which has always cared for children's spirits: the woods."

Very young children love animals—this is true. Early years' books, folktales, nursery rhymes, teddies, emblems on clothes, first noises, first words, smiles, laughs and squeals usually relate to dogs, cats, horses, cows, pigs, birds and other denizens of the animal kingdom. The most common first word apart from "Mama" and "Dada" is "cat." Animals are given personal pronouns in stories, assigning personhood and value and character. And then the linguistic influence changes course and they are taught that a member of a non-human species is an "it." As Robin Wall Kimmerer puts it, we "put a barrier between us, absolving ourselves of moral responsibility and opening the door to exploitation." How would we think of an oak tree if it was referred to as "he" or "she," instead of the inanimate "it"? With a greater sense of communion and relational bond, perhaps.

Then, instead of nurturing that early relationship with the natural world, instilling a deep understanding of how healthy ecosystems work and how to value the natural world for its intrinsic beauty, wonder and mystery, we enclose children inside.

In Britain, by law, schools, nurseries, pre-schools and childcare providers that look after children until the age of five must adhere to the Early Years Foundation Stage (EYFS) statutory framework. Children should have access to outdoor space or daily activities outside, but the framework doesn't say anything about natural areas or wildlife or contact with nature. The "outdoor space" can be fake grass or concrete. There are seven specific areas in the curriculum, and "environment" is mentioned as the final bullet point in "Understanding the world," which "involves guiding children to make sense of their physical world and their community through opportunities to explore, observe and find out about people, places, technology and the environment." Clearly, maths, literacy, art and design and emotional development are crucial, but is it any wonder that children aren't connecting as much with the natural world if it isn't prioritized in their earliest years? It feels like an afterthought because it is an afterthought.

We've known for a while that the presence of greenery affects the quality of play. A 1998 study in Chicago compared children who played in "vegetated" outdoor spaces with "barren" outdoor spaces. Those who lived in the latter played outdoors half as much as those who lived in spaces with more trees and grass. The researchers also reported that creative play was much lower in barren spaces.

"It is necessary to be outside for our brains to be stimulated from the flow of sound, light, shapes and colours that nature provides," said the late David H. Ingvar, an influential and pioneering neurophysiologist at the University of Lund in Sweden, "especially between the ages of three to six, when the energy flow in the human brain is at its greatest." This matters because creative play is particularly key in emotional, social and cognitive development. Also, children in inner-city urban areas with the least access to greenery and vegetation are at a higher risk of social factors such as poverty, poor housing and dangerous neighbourhoods affecting their chances of development. The evidence is clear: children need to spend time in nature.

Numerous studies show that outdoor learning boosts children's social and psychological growth. And yet, amazingly, greenery, trees or flowers are not mandatory or even recommended in the official guidelines for playgrounds or outdoor space in Britain's schools. "Much of society—including much of the education establishment—no longer sees independent, imaginary play, especially in natural settings, as 'enrichment,' " the writer on nature-deficit disorder Richard Louv told me.

In Sweden, a long-time pioneer of children's outdoor education, the use of an evidence-based tool called OPEC (Outdoor Play Environment Categories) demonstrates why green areas have a positive effect on play. It explains the "flow" of an environment which allows children to switch between running, jumping and climbing and "pretend play and contemplative recuperation." Exploring the undiscovered, weaving between trees, navigating corners, using open spaces, transforming a log into a ship, or a stone into a cake is what the OPEC tool is designed to encourage. "A playground with a lot of equipment can house many affordances for physical activity but can still have an overall design that does not support children's dynamic play-flow," says creator Fredrika Mårtensson, an environmental psychologist at the Swedish University of Agricultural Sciences.

The evidence is mounting. In 2007, UNICEF published a paper on child well-being. When children from Spain, Sweden and the United Kingdom were asked what they needed to be happy, the top three answers were time, friendship and the outdoors. Another study tracked people who were in the Scouts or the Guides in childhood and found that they had better mental health in later life. Greater contact with the natural world—as opposed to the simple presence of green space—in childhood was correlated with fewer depressive symptoms later in a study of Australian adults. Studies from Copenhagen, New Zealand and Italy suggest that walking in a natural environment can decrease symptoms of inattention in children with attention deficit disorder/attention deficit hyperactivity disorder (ADD/ADHD). (In the United States, rates of ADHD have increased from 7.8 percent in 2003 to 11 percent in 2011.)

Vulnerable children may need it more than most. Access to nature was found by environmental psychologists Nancy M. Wells and Gary W. Evans of the Cornell College of Human Ecology (2003) to provide a buffer to life stresses in rural children, particularly vulnerable children. Contact with nature seemed to moderate or dampen down the psychological stress caused by events such as bullying at school, the death of a grand-parent, moving house or fighting with parents. The effect was found to be more powerful for the children who were most disadvantaged and subject to the highest number of stressful life events. Nature, in this study, was measured by looking at three elements: one, whether the view from the child's house looked onto trees, plants or other natural elements, a "non-natural" view or no view at all; two, how many plants there were in the living room; and three, the material of the yard, whether it was grass, dirt or concrete. "If access to nearby nature is indeed a protective factor, contributing to the resilience of children and youth, then if nearby nature is lacking, it is one more strike against poor children who already face tremendous disadvantage," they wrote.

Urban living and lack of access to nature could be affecting children in still other, serious ways. A major study of 500,000 children in Sweden found a correlative link between air pollution and mental illness, building on earlier studies that showed the brains of children are particularly vulnerable to air pollution. What is so striking and worrying about the study was that Sweden has relatively low levels of air pollution. The EU and WHO limit for nitrogen dioxide ($NO_2$) is 40 micrograms per cubic metre, which is often exceeded many times over by major cities such as Chongqing, Cairo, Mexico City, New York, Paris, Tokyo and Lahore. The researchers found that an area with just 10 mcg/m$^3$ corresponded to a 9 percent increase in mental illness.

In 2016, as many as 433 schools in London were located in areas that exceeded EU limits for $NO_2$ pollution, and four-fifths of those were in deprived areas. It makes decisions to cut down trees (which reduce particulate matter) in urban areas, such as in Sheffield, Newcastle and Edinburgh, look even more foolish.

FEW STUDIES FOCUS on adolescent mental health and nature exposure, but some are starting to be published, which is prescient considering the rising levels of self-harm, mental health disorders and rates of suicide among teenagers. In 2018, a team at the Harvard T. H. Chan School of Public Health found that surrounding greenness was linked with lower odds of depressive symptoms in a study population of 9,385 American teenagers. To investigate further, they conducted a study with longitudinal data, following those adolescents and young adults aged twelve to eighteen. Those who were exposed to higher levels of nature (or vegetative density, as the study puts it) during childhood and adolescence had a lower risk of depressive symptoms in later adolescence and also adulthood. The effects were strongest for those living in areas of higher population density. We now know that air pollution particles can cross to the foetal side of the placenta, so that babies are affected before they are even born.

In 2019, a team at King's College London found a significant association between adolescents living in polluted areas and reports of psychotic experiences.

Something is clearly rotten in the state of childhood and adolescence. Thousands of children in Britain are self-harming and facing stress and mental illness, with teachers constantly raising the alarm that schools are facing a health crisis of intolerable proportions, created by a combination of social media, pressure on exams, and austerity measures which lead to an absence of adequate healthcare resources. Judging by the evidence, a loss of connection with the natural world is a factor. Are we forgetting to let children revel in the peace of wild things and the softness of the dirt? As the epidemiologist Howard Frumkin argues, "Nature contact is not just an amenity; it is a birthright."

Accomplishing deep cultural change doesn't happen overnight, but there is an exciting movement brewing: if you go down to the woods today, you might just bump into a class of children. In Britain, a growing number of outdoor nurseries are opening, and teachers are taking their classes to natural areas to learn, to play and to imagine.

IT WAS EARLY, not yet nine o'clock, but the July sun was in full array. Through a gate, a group of eleven children, aged between two and four, rushed into a vast field. The meadow pulsed with crickets, and a lilac haze of grasses stretched to the horizon. A couple of four-year-old girls pored over a wildflower identification sheet, worked out that the bright yellow plant was Lady's bedstraw and recorded it in their notebooks. A young toddler called Wilbur stood to one side, blowing the seeds from a dandelion clock. The children looked for insects. Suddenly, an eruption of delighted giggles: a grasshopper had been found.

We walked across the field to a large clearing in the woods shaded by oak trees. The "staff room" was a shack decorated with tiny animal skulls, pine cones and fluorescent backpacks. At the other end of the space, there was a toilet (compost) and a small hut (stylish) for use in particularly cold or wet weather. The village hall was available for days with severe storms or gales, but the children were outside most of the time.

The founder and owner of outdoor nursery Elves & Fairies, Kirsteen Freer, had long, wavy apricot-coloured hair, a calm presence and very good eye contact. She wore a wide-brimmed, elf-like hat and a flowing tunic. Inspired by her own childhood of camping, picnics and walks, she raised her children on handiwork, woodwork and cooking instead of watching television. In the holidays, she sent them to Forest School camps. "They didn't become grumpy, sullen teenagers in the way lots do," she said. Instead, they remained chatty, interested in life and rarely bored, which she chalks down to a childhood spent outdoors. "I realized that there was an awful lot of children who didn't have those experiences," she said. So, while teaching piano and folk-dancing at a Waldorf school (establishments that prioritize creativity and imagination), she decided to start a kindergarten for children who wouldn't be able to afford a Waldorf education.

In 2007, Elves & Fairies opened with just two children, and she struggled to increase the numbers. In the first few years, parents already following a more alternative way of life drove in from miles away, but it took time to convince local families that it wasn't just a "hippie pre-school." "I think they felt that if the children were just playing outside, how were they going to learn anything?" said Kirsteen, who smiled and laughed when she spoke, but clearly had a focused passion for holistic, outdoor education.

In January 2017, the local perception changed when the Office for Standards in Education, Children's Services and Skills awarded the nursery an "Outstanding" rating. "All of a sudden the local people realized that 'Oh, wow, that strange school is actually functioning as something worth sending our children to.'" From then on she was inundated with interest.

Elves & Fairies offers the same curriculum as other nurseries. However, the technique is different. It is experiential, child-led and outdoors. Children learn maths through counting molehills, sticks and bowls for lunch; "understanding the world" works by noticing seasonal changes, taking small risks on a rope swing and learning how to make a fire. Emotional development takes place through increased resilience to mixed weather conditions, and physical development through climbing logs or using a knife to cut vegetables. It is simple and old-fashioned; sticks and mud instead of toys and iPads.

The morning begins with free play, crafts, gardening, singing or watering the plants. Everyone helps with lunch: finding sticks, laying the fire, chopping the vegetables, cracking eggs. Then, it's nap time for some—a couple of little ones snoozed happily outside when I visited—or stories, sewing, more free play and reading.

One of the parents, Kathryn Footner, had sent her three children to Elves & Fairies, attracted by the natural, nurturing atmosphere. "I like the simplicity of the toys, the fact that instead of providing a kitchen for the children to play with, they use logs and pans and the children use their imagination," she said. Eden, her five-year-old, had cerebral palsy and the setting suited her needs. To stimulate parts of Eden's brain, Kathryn varied her environment and activities. "Nature provided the best variation in the most organic way. The plants change, the sky changes, the ground changes, the length of the grass."

Gemma West, another parent, was attracted by the environment and lack of screens. Max, her two-year-old, attended and she attributed improvements in his speech to the nursery. Instead of worrying how he might transition into a classroom setting, she considers it the opposite. "He's a lot more social than my other two boys."

Sausages sputtered and hissed on the fire and the smoke billowed around the camp. Without complaint, the children helped prepare lunch: some fetched and carried the plates, others grated cheese or watered the vegetable garden for future meals. Then, suddenly, there was a commotion around the fire. Verity, a warm and inventive teacher, had found a coal fungus, or King Alfred's cake. The eerie black rock was passed around in awe. Inside its matte chambers, there was a moist, white larva. "What's the larva going to turn into?" "A little baby dinosaur!" The children gathered to see it with a magnifying glass and discuss how the tunnels were made. It felt very calm and peaceful, with no crying or fighting or meltdowns. Kirsteen put this down to the environment. "Nature is calming," she said. "The children have got more room and space, they're not overcrowded, they're not surrounded by noise, or overstimulated."

"Children don't need that much," she said, referring to toys, games and other objects to facilitate play. "I think people need to start looking back a little bit at what children used to have, yet they were still capable of learning. What we're offering is a childhood that our grandparents would have had. Nature played such a big part for children in the past. After breakfast and before school, they'd go outside . . . For very sensitive children, being inside can be a big challenge. Outside they've got space, it's not so noisy, and it's the same for the teachers."

The setting of Elves & Fairies is relatively unusual in its complexity and lushness. A private landowner allows the nursery to use her wood-land. Most children and parents in Britain or early years care-providers don't have access to a garden, let alone a beautiful woodland across a pic-turesque meadow. However, this hasn't stopped the drive to reconnect children with the natural world, even those in urban areas. Go to Kew Gardens or a park in other British cities and you might well spot a huddle of toddlers or young children in high-vis vests, looking for worms under logs or examining tree bark. Currently, these youngsters are examples of the one in four children who spend more time outdoors than prisoners, but the movement is growing.

ONCE CHILDREN START SCHOOL, they can begin "forest school," which refers to regular, frequent, child-led, risk-based sessions of learn-ing and play in a natural area, ideally a woodland, rather than an actual school in a forest. Schools may incorporate it into their curriculum or children will attend in holidays and half-terms. Since the first forest school was set up in 1993 at Bridgwater College in Somerset, inspired by a visit to a *skovbørnehave* (Danish forest kindergarten), the idea took root in Worcestershire, Oxfordshire, Shropshire, Norfolk and Devon, before spreading into schools across the country.

I was out of primary school by that point and can't recall any outdoor learning or forest school at secondary. I loved learning about the layers of a rainforest—the canopy, dripping and steaming, the understory, man-groves and swamps, giant ferns and tree trunks like witches' hands, the forest floor—in Geography, but I was discouraged from continuing with the subject because, I think, it wasn't seen as a "proper" A-level in certain academic quarters, which again illustrates our disconnection from the world in which we live.

By 2017, forest school was on the move. Practitioners had trained teachers in every county in England. It's catching on overseas, too: forest school leaders were trained in Dubai, Australia, Budapest, China, the United States and Amsterdam in 2018.

As concern grows in Britain about the number of children being treated for anxiety and depression, amid evidence of the adverse consequences of a lack of time spent in the natural world, the forest schools movement is a light in the darkness. Among parents, there is an appetite for "greener" learning, with "nature schools" now starting up. In May 2017, a petition calling for a GCSE (General Certificate of Secondary Education) in natural history was signed by more than ten thousand people. "Re-engagement with Britain's natural history has never been more urgent. Young people need the skills to name, observe, monitor and record wildlife. It is vital to understand the contribution nature makes to our lives physically, culturally, emotionally and scientifically both in the past and today," it said.

The UK government, which was Conservative at the time, responded dismissively: "On the basis that natural history subject content is covered in the current national curriculum and in a number of GCSEs, and given the thorough and careful work that has gone into developing, designing and approving reformed qualifications, the Government does not currently intend to introduce an additional GCSE in natural history." Its priority, it said, was "to give schools time and space to provide excellent and inspiring teaching of new qualifications." There was no acknowledgement of the need to understand the natural world in order to look after it, nor of the benefits of a relationship with nature. It is true that natural history is incorporated into other subjects, such as Biology and Geography, but, really, how can you truly learn about the natural world, about habitats and ecosystems and how plants photosynthesize, without seeing them up close?

At a time of rising mental health problems in primary and secondary school-age children, there is evidence that forest schools can foster resilience which may help to reduce stress. Over a four-year period of a study on outdoor learning, delivered through Plymouth University, 90 percent of pupils from 125 schools said they felt happier and healthier after learning outside.

When I talk to teachers about the benefits they have observed, the word most commonly repeated is "confidence." Through outdoor play, the children, says Kirsteen Freer, grow in independence, both physically and mentally. Studies of the long-term impact of forest school back up her claim, strongly suggesting that time in nature can help children build confidence and self-esteem. A class that hadn't gelled at the James Dixon Primary School in East London was prescribed forest school for a term. "It stopped low-level disruption and helped with compromise and resilience," said teacher Fred Banks.

Considering the benefits, it would be wonderful if every child at school in Britain had the opportunity to learn outside and attend forest school. However, there are problems: not all teachers want to go outside in the dirt and the mud. Our severance from nature is so severe, how many teachers would be keen on outdoor learning, given the choice? Government and schools should do everything they can, however, to make it possible, especially in urban areas where there are even fewer opportunities for children to see green and natural spaces or wildlife. Some children in urban areas will only go outside on their way to and from school.

WALK INTO Tower Hamlets Cemetery Park (THCP) in East London, and instantly the smell of car exhaust is exchanged for air that smells of mulch and musk. Walk a little further into these thirty-six acres and there's a whiff of horse manure, to feed the plants. The gravestones are higgledy-piggledy, cracked and overgrown with moss and ivy and other plant life. Squirrels chatter, pigeons warble and a magpie makes the sound of a wind-up toy. This spot is a wild oasis in the heart of London, and a venue for forest school in term-time and the school holidays. On the last day of May, I observed a half-term session.

Through the trees, there's Canary Wharf and the bright livery of the overground train. An aeroplane rumbled overhead as we cooled down under a large sycamore tree. There were thirteen children, aged from three to seven, and the activities were riskier than at Elves & Fairies. A group used parabolic mirrors to make a fire. Some whittled sticks with sharp knives, others hammered wet charcoal. After ninety minutes, they explored deeper into the woods, the edges of exploration marked out with stop signs. In one corner of the woodland area was a mud kitchen, in another a clay area, where children made totem poles or creatures. Some daubed their little faces in camouflage paint and hid behind large fronds of bracken. "Can you see me? Can you see me?"

On the train up to London that morning I'd read about a survey by the Royal Horticultural Society that found we barely know the names of plants any more. Half of adults can't name a shrub and four out of ten can't name any houseplants. I left the forest school that day with more knowledge. Amelia, aged six, taught me, patiently, that the juice of a nettle is better at soothing stings than a dock leaf, that the butterfly we saw was a holly blue and that a stag beetle I spotted was actually a lesser stag beetle. She also put me straight on who would win in a battle between a centipede and an ostrich.

Kenneth Greenway, the park ranger who ran the forest school, had a boyish face, an earring and a wide smile. He wore a racing green–coloured uniform with an ATHEIST badge pinned to his lapel. First off, he set out the rules while we sat in a log circle. "Don't fight with sticks, they are for imaginative play. No boy/girl rubbish, no dividing the genders. I don't like it. Be kind and considerate. Don't stick things in your mouth if you don't know what they are."

Bug pots were passed around for quick-moving insects and identification with ID sheets at the bug station. A ground beetle! A ground beetle! He's racing across the sheet! Greenway put it in the glass pot and sent it around the circle so we could have a closer look. It looked as if it was made from charcoal, striated across its elytra and smooth thorax. The legs were a molasses brown and its antennae waved here and there, sniffing around.

"Do you want to go and find butterflies with me?" asked Amelia. We got sidetracked, overturning logs to find slugs and woodlice. "Bandies! Bandies!" her little sister Elsie, who was four, cried. "Snails are so kind. They are friendly," she mused. We picked up worms to observe the way they slink. "The only problem with being outside is that you can't see the iPad screen because of the light," a boy, five, told me later.

Tower Hamlets is an area of London with minimal green space, a high proportion of ethnic minority children and high poverty rates. Free and subsidized places are offered at the THCP forest school to ensure its inclusivity. In the United Kingdom, people from ethnic minority groups are less likely to access high-quality natural areas compared with the rest of the population. This can potentially exacerbate health inequalities in children, as well as in adolescents and adults, as I will explain further in Chapter 6, Equigenesis.

In June 2018 a study, the first of its kind, was published in *The Lancet Planetary Health,* assessing the contribution of green space on the mental health of children. It looked at a multi-ethnic group of four-year-old children in a highly deprived urban area of Bradford. It examined the availability of green space, how it was used and how it was perceived. The children were predominantly of South Asian origin (58 percent) and the rest were of white British origin (29 percent) or another ethnicity (13 percent). The study had robust controls for demographics, socioeconomic status and maternal health behaviours. It found that greater satisfaction with a green space was independently predictive of the South Asian children's mental well-being. "This is an important finding showing that quality may be as—if not more—important than quantity in mediating green space and health associations, and might be particularly significant in multicultural contexts," wrote world design and health expert Jenny Roe, in a commentary on the study.

Greenway, the park ranger, told me about children who were initially reluctant to engage with the woodland, or anything that might be dirty or dangerous. Some of the children at the sessions were brought in as a class from schools in the local area. "Then they become enthused by it," he said. "They want to do bug hunting. They want to do muddy play. You see a change in play behaviour. You see excitement in children when they see a beetle ... They go from being cautious, trepidatious, with a lot of worry, and they come back happy, excited, just from being outdoors. Teachers also say, 'I feel really good, I've really enjoyed being in the woods, it smells different, hearing the birds.'"

Children today have vastly reduced opportunities to connect with nature compared with fifty years ago, and forest schools offer a formalized, thought-through, well-evidenced way of incorporating a relationship with the non-human world into education. Children spend most of their time at school, so the responsibility for providing opportunities for a connection with nature must lie at least in part with educators. Most schools don't have grounds or playgrounds with rich biodiversity, so the move to take classes out to forests and woodlands for lessons and activities is sensible and, considering the evidence for health and well-being, critical. It is one solution to the problem. By cooping children up in classrooms for thousands of hours, we are doing them a disservice. Children should be spending time outdoors in order to foster the relationship and develop socially, as members of a community, and of the wider community of Earth. The framework for forest school is already there and available, but we need the political will to see the potential for children and society. Teachers in Britain are currently so over-stretched that it will require deep cultural change to build proper environmental and outdoor education into our school system. But it is a matter of public health at a time when children are suffering. And without introducing children to the interconnectedness of life, the gushy, gross wonder of the roiling, seething, slimy, dirty aliveness of nature, how will they love it, and how will they protect it? The possibilities for us, our children and our society, if we have a renewed connection with the outside world, are too huge—and urgent—to ignore. The dangers for our children and society if we continue to lead them away from the natural world and pretend that everything is OK are too frightening to comprehend.

Looking back over the months that followed, I recall my afternoon at the London forest school as a brief dip into wonder. Lying belly-down beside the pond and looking for newts with two young girls, who taught me how to tell the males apart from females, I felt awed by the strangeness of the natural world and reminded of my own childhood, when stopping to observe the perfect rings on a pink worm, or the flicking eye of a lizard or the matte tusks of a stag beetle was so exciting I felt I would explode. Often I forget these experiences of awe and enchantment are still just around the corner, for all of us, if we would only look—and their effect on us is no minor thing.

PART III

# BRANCHES

∽

# 4

# Physiological Resonance

Whoa, that's a full rainbow all the way. Double rainbow, oh my god.
It's a double rainbow, all the way. Whoa that's so intense. Whoa man!
Wow! Whoa! Whoa! Whoa ho ho oh my god! Oh my god! Oh my god!
Woo! Oh wow! Woo! Yeah! Oh ho ho! Oh my god! Oh my god look at
that! It's starting even to look like a triple rainbow! Oh my god it's full
on! Double rainbow all the way across the sky! Oh my god. Oh my god.
Oh god. What does this mean? Oh. Oh my god. Oh. Oh. God. It's so
bright, oh my god it's so bright and vivid! Oh. Ah! Ah! It's so beautiful!
[Crying? Laughing?] [Pretty sure he's crying.] [Now he's laughing and
crying.] Oh my god. Oh my god. Oh my god! Oh my god, it's a double
complete rainbow! Oh right in my front yard. [Laughter] Oh my god.
Oh my god, what does it mean? Tell me. [Crying] Too much. I don't
know what it means. [Laughter] [Heavy breathing] Oh my god it's so
intense. Oh. Oh. Oh my god.

> —Transcript of the viral Double Rainbow 1–8–10 video
> by Paul "Bear" Vasquez, Yosemitebear62

In those dark days I found some support in the steady progress
unchanged of the beauty of the seasons. Every year, as spring came
back unfailing and unfaltering, the leaves came out with the same tender green, the birds sang, the flowers came up and opened, and I felt
that a great power of Nature for beauty was not affected by the war. It
was like a great sanctuary into which we could go and find refuge.

> —SIR EDWARD GREY, Foreign Secretary
> at the outbreak of the First World War, from
> "Recreation," speech to Harvard University, 1919

S O MANY OF US emerge from childhood with a degraded relationship to the natural world. We are taught that we are separate from the land and the earth and the other creatures that dwell in it. Instead of being awed and intrigued by the delicious grossness of, say, a worm, or the spider that eats its mate, or the behaviour of the cuckoo, we are inclined to recoil and feel a simple disgust for other animals, especially those with whom we share our urban areas: the foxes and rodents and wasps that are cast as vermin or pests in our anthropocentric imaginings.

We forget what we have learned in that boring biology lesson in a dusty old lab about photosynthesis and the interconnectedness of life. But in doing so we are overlooking a relationship that can offer our minds, and even our brains, much advantage.

As I shifted from childhood to adolescence, I lost my early connection with nature—and that regular feeling of being wowed by the outside world. I imagine this is common at a time when socialization, making friends, testing boundaries and identity-creation become of primary importance. Being earnestly struck by wonder as a teenager in my grungy, indie pocket of the 1990s was hardly very cool. The transition to adulthood was a turbulent time for me, as it is for many, when patterns of self-destructive behaviour took root, and I experienced physical symptoms of stress and emotional discomfort, such as chronic eczema and shingles. Of course, the biological and hormonal upheaval, as well as the interpersonal Sturm und Drang, would have still been there, but I wonder if my alienation from nature was also a factor, considering how important it was in my recovery in my late twenties. It certainly didn't help.

One of the aspects we lose without a relationship with the rest of nature is regular opportunities for awe. We might associate awe and wonder more with the naivety and innocence of childhood—wow, says the child who sees an octopus or snow or fireworks for the first time—but new research suggests it's far more potent in adult life than we might think. A new area of research—the science of awe—explores the specifics of how natural phenomena make us feel, and could have significant implications for our mental and psychological health, and even how nice we are to other people.

Before the Irish philosopher Edmund Burke revolutionized the concept of awe in his *A Philosophical Enquiry into the Origin of Our Ideas of the Sublime and Beautiful* in 1757, awe was mainly ring-fenced for the religious and the sacred. Awe-inspiring experiences of nature, such as visiting the Grand Canyon or white-water rafting, were not available for the majority of the human population, then or now. Is awe a luxury today? Or, as John Muir, the American-Scottish naturalist, writer and advocate for the wilderness, put it, does everyone need beauty as well as bread?

In the 1990s, a psychologist called Dacher Keltner at the University of California at Berkeley took a scientific approach to the emotion of awe for the first time. Many experiences of awe in the modern world still come from an encounter with nature, despite our disconnection. Perhaps unsurprisingly, Keltner found that awe increases happiness and lowers stress. He and his researchers found that awe is actually a common experience: people tend to experience awe on average two and a half times a week, and some groups feel awe more than others. Older people experience more awe, and so do women. You may even be a carrier of a dopamine-related genetic polymorphism that directs how much awe you can experience.

Why the experience of awe evolved was an early interest for Keltner's lab. One evolutionary theory proposed that pre-cultural, primordial awe related to the emotional reaction of a subordinate to a charismatic leader, which helped create the hierarchical structures that allow societies to function and survive in uncertain environments. "Awe binds us to social collectives and enables us to act in more collaborative ways that enable strong groups, thus improving our odds for survival," wrote Keltner.

This idea was borne out by some fascinating studies in how experiences of awe affect the body and the mind. For example, an intriguing study about cytokines (an overactive cytokine response is associated with disease, depression and ill health) suggests that awe has an important physiological effect. Professor Jennifer Stellar, from the Department of Psychology at the University of Toronto, Mississauga, measured cytokine levels in samples of gum and cheek tissue and the presence of positive emotions in two hundred young adults, while she was working in Keltner's lab. She found that only awe predicted reduced levels of cytokines—Interleukin 6, an inflammation biomarker—to a significant degree. "That awe, wonder and beauty promote healthier levels of cytokines suggests that the things we do to experience these emotions—a walk in nature, losing oneself in music, beholding art—have a direct influence upon health and life expectancy," said Keltner, who was co-author of the study. And it may even affect people suffering from mental illness. After whitewater rafting, PTSD symptoms of military veterans decreased by 30 percent, and they reported less stress and an increased sense of well-being.

It may also affect how we behave and treat each other. To see how awe might change perceptions of self and of other people, Keltner showed one group of participants a video of canyons, mountains and other awe-inspiring scenery, and another group a natural scene that was supposed to be funny. Afterwards, both groups were told they'd won a prize. They were then asked if they wanted to share their cash prize with strangers. The people who'd laughed at the comedy scene were a lot less keen. They wanted to keep their winnings. But the people in the "awe" group were more likely to share winnings from a lottery cash prize with strangers afterwards. People were more ethical, kind and generous after feeling awe—and this phenomenon has been replicated in experiments over and over again.

Why would people be more generous and kind after experiencing awe? Perhaps, in part, simply from being in a good mood. Looking at the brains of those in the groups at the time gave the research team a clue as to what might be happening. Using functional MRI, scientists saw that awe reduced activity in the default mode network, the area of the brain associated with the sense of self.

Awe, then, can shift us away from pure self-interest to be interested in others. It can help us bond and relate to each other. It can turn off the self, the day-to-day concerns, to propel us into focusing on something bigger and hard to comprehend. It reminded me of a phenomenon I'd heard about a lot when spending time with other people recovering from drug and alcohol addiction: the shutting off of the self, the turning down of "addiction.fm" or the "washing-machine head," the ruminating thought processes that a substance could hush or turn down temporarily but, in the long run, will feed. Awe may even be an antidote or counterpoint to the narcissism that people are worried social media encourages today.

There is an abundance of wonder in our home that we are losing as habitats shrink and our connection wanes. Antlers. Orcas. Seastars. Stag beetles. Curlews. Rotifers. Toadstools. Glow worms. Puffins. Bats. Chrysalides. Shooting stars. Red velvet mites. The rosy maple moth. Christmas tree worms. The bioluminescent strawberry squid. Pygmy shrews. Wolves. Crows. Opals. Nudibranchs. Owls. Really, *awe* is Earth's signature. We may have forgotten, but how could it not be? The adorable, terrible, leaky, stinky, gooey, glimmery, furry, bloody, swoony, shimmery, thumping majesty of the Earth. The very earthiness of it. It claws. It kicks. It rots. It ruts. It squawks. It squeals. It chomps. It bursts. What a wild and mind-bending disco there is on the Earth, if we would only look and take notice!

WHAT DO WE MEAN, exactly, when we talk about good mental health and well-being? Calmness? Communion? Bliss? A mind joyous and free? The absence of terror or obsession or pain? One aspect of mental health that has been researched in the context of nature connection is restoration, especially in conjunction with mental fatigue.

One day in June, I walked towards a secluded cove in Galloway, Scotland. I passed orchids, cream blossom, cow parsley, bluebells and vetch. I could smell the salt of the sea and the trees, the splendour of spring broiling in the sun. I stopped to watch a lurid greenfinch before it retreated deep into a gorse bush, and I made the descent down a grassy rockface, gripping clumps of grass and trying to avoid the wildflowers, the "eggs and bacon" (bird's-foot trefoil) and the carnivorous butterwort. Nestled between rocks sat a trio of perfect oystercatcher eggs. Ringed plover were nearby, scurrying officiously, guarding their cargo. Here and there, fritillary butterflies twisted and turned about each other, like the whirling tails of a kite. The landscape was pulsing with life and my heart beat faster in response.

White sand, buffed glass and mussel shells provided a carpet until the rock fissured into a narrow corridor which opened to the ocean. The sea was mirror calm, transparent and cold. I peeled off my swimming costume to feel it even closer on my skin. Lying on my back in an area hit by a ladder of sunlight, I heard a chirruping and something swooped towards me. I saw from its beak, a long Plasticine carrot, that it was an oystercatcher. Two Maleficent-like cormorants flew past higher up. I stayed for a while, treading water, licking the salt from my lips, relishing the solitude. I felt part of the earth, the rocks and the ocean. It was now; it was cold; it was little else. At the time, I'd been sober for a couple of years, but I still felt my alcoholism acutely. My head was swarming with doubt and uncertainty, anger and resentment, frustration and boredom. After the swim, my brain felt scoured clean.

It is this sensation—feeling energized, calmed and restored by being engaged, effortlessly—that is conveyed by the term Attention Restoration Theory, or ART. In 1980, Rachel and Stephen Kaplan, professors in psychology at the University of Michigan, came up with the concept of ART. They set out the idea that too much attention on one thing (Directed Attention) could lead to mental fatigue (Directed Attention Fatigue), which could cause impatience, stress, irritability and trouble concentrating on a task. Spending time in a natural setting is ideal for Effortless Attention, they posited, which effects a state of Restored Attention, achieved through "soft fascination" (leaves moving in a tree, a river flowing, birds in the sky), "being away," compatibility (the person wanting to be there) and extent (the scope for immersion in a landscape or environment).

The theoretical basis for ART was the writing and thinking of the nineteenth-century American psychologist and philosopher William James (brother of novelist Henry James). In *The Principles of Psychology* (1890) James explained that there are two types of attention. One, "Involuntary Attention," means attention that requires no effort, the noticing that is unconscious, automatic and instinctive: for example, when a young deer jumps in front of you or your attention is drawn by a sailing boat on the horizon. The second type of attention is "Voluntary," and takes effort and concentration. James thought Involuntary Attention gave the brain a rest so that it could engage in Voluntary Attention when it needed to. The Kaplans took this one step further, suggesting that an imbalance between the two states could lead to harmful mental fatigue: feeling irritable and distracted and making impulsive decisions. The natural world, they suggested, is an ideal tonic for such fatigue. "If you can find an environment where the attention is automatic, you allow directed attention to rest. And that means an environment that's strong on fascination."

When ART was first suggested, it was largely descriptive. In the last thirty-five to forty years, scientists have tested it in different ways. An experiment published in the *Journal of Environmental Psychology* in 2015 compared the effect of "micro-breaks"—viewing either a flowering-meadow green roof in a city, or a bare concrete roof—on cognitive functioning. Participants were given a task to complete, then asked to look outside for forty seconds, then given the task again. Those who looked out at the green roof made fewer errors and achieved better results. The scientists concluded that the study extended Attention Restoration Theory and suggested that the green micro-break boosted subcortical arousal and cortical attention control. The study author Dr. Kate Lee said that, even though we might not realize it, the reason we look out of a window to stare at a tree might be because it helps concentration if we're feeling mentally fatigued. The study also suggests that it would be in the interests of companies and workplaces to place trees outside windows, and potted plants in conference rooms, for both the well-being and performance of their employees.

Another group who may benefit from attention restoration in a natural environment is children with ADHD. In 2009, a study from researchers at the University of Illinois measured the concentration of children with ADHD after a walk in a green city park, a walk in a downtown area, and a neighbourhood walk. The team found that twenty minutes in the park was sufficient to improve attention performance compared with the same amount of time at the other settings. The authors concluded that "'doses of nature' might serve as a safe, inexpensive, widely accessible new tool in the toolkit for managing ADHD symptoms." For adults—working, juggling tasks, busy and stressed in our modern-day, 24/7, "always on" culture—restored attention sounds like something many of us could benefit from.

WHAT WAS ACTUALLY HAPPENING in my brain during that walk down to the sea? What do we know about how the brain is affected by contact with nature? The last quarter-century of brain-imaging techniques have allowed scientists and medical practitioners to discover remarkable things about how the brain is arranged, which parts do what, and how damage to a certain part might cause disease or psychiatric conditions. Magnetic Resonance Imaging (MRI) assesses the amount of blood-flow in the brain to see which parts are active, in the same way that Dacher Keltner's lab did. Electroencephalography (EEG) measures the electrical activity of the brain using electrodes placed on the scalp. We know that the temporal lobe plays a role in emotion and memory, and that the hypothalamus controls the pituitary gland, which manages our hormones. We know about neurotransmitters and how they work, and that dopamine is associated with the experience of pleasure and serotonin relates to mood, sleep and temperature. But there is a lot we still have to learn about the brain as a whole, and the brain in relation to its environment. Both psychology and neuroscience have hitherto tended to neglect the relationship of the human brain and mind to the natural world.

However, over the last decade neuroscientists have started to explain why many of us feel better after a walk in the woods. Cerebral activity in the prefrontal area of the brain, the part involved in executive functions, decision-making and other complex cognitive processes, may be reduced. Your body may produce lower levels of cortisol, which is released to help the body respond to stress. There may also be reduced activity in your subgenual prefrontal cortex, a small area in the cerebral cortex which is associated with sadness and negative rumination or brooding. When a person is in a natural area, rather than an urban, non-natural environment, their brain tends to be less stressed, which in turn leads to better mental health.

It may have something to do with the benefits of a stimulating environment. In 1947, Donald O. Hebb, a Canadian neuropsychologist, demonstrated that exposure to an enriched, multi-sensory, stimulating environment improved both motor and cognitive function in rats. Animals that were taken out of their small cages and given a run-around in his family home for a few weeks were better at problem-solving and memory tasks than those that had remained in their cages. Hebb's research led to a scientific model called "environmental enrichment."

In the 1960s, a research team at the University of California, Berkeley, discovered that an enriched environment actually altered the structure of the brain. The group, led by the American research psychologist Mark R. Rosenzweig, ran an experiment. Two sibling rats were given the same food, light and heat, but different living conditions. One had mud, a wheel to run up, a rope to saunter across, bits and pieces of junk to explore and other rat friends. The other one had just food and water; it was kept in an empty cage with nothing to look at, play with or, well—who really knows?—think about. The results of brain scans showed that the poor rat with the empty home had a quarter fewer synapses than her sister. Her cerebral cortex—the folded, crumpled mantle that forms the brain's outer layer where thoughts, memories, perceptions, consciousness are formed— was also thinner, by 7 percent. A lack of visual or physical stimulation led to a synaptically impoverished brain. Brains become stronger the more they are used. And a healthy, strong, stimulated brain means improved cognition, function and memory.

The scientists didn't fill the lucky rat's cage with palm trees, butterflies and a flowing river, but the results showed a causal link between the level of stimulation in an environment and brain activity, size, complexity and development. Scientists hypothesized that higher levels of multi-sensory stimulation might trigger molecular cascades which activate neurons (the nerve cells in the body) and neuroplasticity (the brain's ability to adapt, reorganize and adjust). Another concept, the "inoculation stress hypothesis," suggests that a stimulating environment may expose an organism— animal or human—to mild stressors that it then learns to adapt to and develop coping strategies for, which increases resilience and inoculates against further stress. Out walking in nature, for example, the body is responding to temperature changes, the presence of a stone in the middle of a path, perhaps, or other people or species, which, in the past, may have been predators. These mild stressors trigger the nervous system to release different types of molecules into the bloodstream and change the chemical make-up of the body. Whichever the underlying mechanism or mechanisms turns out to be, the research suggests we are overlooking the benefits of an enriched environment for the health of our brains, and one easy way to experience such an environment is to leave the office or the living room and walk into a natural space.

An experimental psychologist might put this down to "arousal theory," which is another way of understanding why a natural environment might be good for us. Arousal theory was developed by Daniel Berlyne in the 1960s and set out in his 1971 book *Aesthetics and Psychobiology*. Every individual has an optimal arousal level, he thought, and the presence of complexity, novelty, uncertainty and conflict—a stimulating environment—leads to arousal and the activation of excitement or pleasure in the brain. Simply put, variety makes people and animals happy. Although he was primarily interested in how art creates an arousal response in people, environmental psychologists refer to Berlyne's theory to explain why people tend to prefer a stimulating natural environment over a bare or built landscape. Unity in variety, if you will.

But as multi-sensory stimulation is important for a healthy brain, wouldn't it follow, then, that urban living, with all its noises and colours, neon lights and new faces, would be good for us? A fast-paced, highly populated concrete jungle is also a highly sensory and stimulating environment, much more so than some natural environments—a view over a monoculture field, say, or a lawned community green space with little biodiversity. In fact, what we do know about living in cities is that it increases our risk of mental health problems. According to a 2014 report from the OECD, based on the results of a project conducted between 2010 and 2013, cities have both health risks and benefits (such as accessible health care and social support), but mood and anxiety disorders are more prevalent in city-dwellers, and the incidence of schizophrenia is much higher in people born and raised in cities. Living in greener neighbourhoods is associated with slower cognitive decline in elderly people.

A neuroscientific approach as to why this might be is in its infancy, but in 2011 a group of German psychiatrists and neuroscientists published a study that showed what city life does to the brain. Thirty-two subjects who lived in a mixture of rural and urban areas in Germany were given difficult mental arithmetic problems to solve, with a short time limit that led to a high failure rate; then they were scolded and hurried via headphones and told that their scores were worse than the average. The study was designed to make the subjects feel anxious about how they would do, using the Montreal Imaging Stress Task, an easy way of inducing a stress response in a human, similar to the Trier Social Stress Test mentioned in Chapter 1. Using MRI scanning, the research team looked at the brain activity of the different groups as they processed social stress.

Only the participants who lived in cities had increased activity in the amygdala, the area of the brain associated with emotions, anxiety disorders and depression. "We know what the amygdala does; it's the danger-sensor of the brain and is therefore linked to anxiety and depression," said lead author Professor Andreas Meyer-Lindenberg of the University of Heidelberg. Those born in cities also had excessive activity in the cingulate cortex, an area that controls emotion and regulates the activity of the amygdala. It is a key part of the limbic system implicated in processing and regulating stress. The study showed that specific brain areas in these subjects were highly stressed by city life. The authors suggested it could be due to noise, overcrowding, pollution or social and demographic issues. "There's prior evidence that if someone invades your personal space, comes too close to you, it's exactly that amygdala-cingulate circuit that gets [switched on] so it could be something as simple as density," said Meyer-Lindenberg. City life can make people more vulnerable to social stress, which is a risk factor for mental illness.

It is a fine balance, however. On the one hand, as the animal experiments and scientific models of environmental enrichment tell us, novelty and social contact is good for the brain. It makes it stronger and more malleable, which translates to better mental health. On the other hand, those living in towns and cities with high population densities might have too much stimulation, to the point where they over-stress parts of the brain which are linked to poor mental health, and fewer opportunities for restoration through natural settings.

As most of us live in urban environments, and need to travel on public transport with lots of other people in order to get to work, how do we strike a balance? Can nature within cities provide a buffer? A group in Edinburgh used EEG kits in a study of people over the age of sixty-five to assess the impact of the urban environment on brain activity while walking between a busy urban space—a commercial street—and a green space—a public park. In both groups, whether they started in the park or the road, they had a high stress response. What was interesting was how green space seemed to have a buffering effect on the stresses of the urban area. The group who started in a green space and walked into a busy, built-up space experienced an increase in alpha brain waves—the electrical activity of the brain associated with relaxation. Sometimes they'd only spend five minutes in a green area, but there was still a demonstrable effect on cortical brain activity. Nature seemed to undo the stress of the city walk, in the moment.

ANOTHER WAY THAT NATURE can affect the brain is through something very simple: shapes. Ferns, seashells, lightning, salt flats, snowflakes, pineapple, fractal broccoli, ocean waves, clouds and mountain goat horns are examples of natural phenomena with fractal features. Fractal, in this context, means a self-repeating pattern of a shape that varies in scale, rather than being repeated exactly. If you can see a tree outside your window, you are looking at a fractal shape.

And there's a related reason why it might be pleasurable or restorative to look at. In the 2000s, Richard Taylor, Professor of Physics, Psychology and Art, and Director of the Materials Science Institute at the University of Oregon, used both EEG and functional MRI techniques to observe brain activity while subjects were viewing different types of fractals, such as Jackson Pollock paintings or snowflakes. Taylor discovered that patterns with a fractal dimension of 1:3 (most fractals in nature fall between the 1:3 and 1:5 interval) provoked high alpha waves in the frontal lobes and high beta waves in the parietal area, suggesting a relaxed but focused state which Taylor argued could reduce stress levels.

Why would we be "hard-wired" to respond to fractals in nature? It turns out that the structure of the eye is fractal in itself, and when it views a fractal image, it locks into place, so to speak. Taylor called this "physiological resonance." Therefore, fractal-like phenomena are easier on the eye than, say, the shape of a toaster or a traffic light, surely because our eyes have evolved to decipher and understand fractal patterns in the natural world.

Looking at fractals in nature is another salve to our minds that, in the face of rising stress levels and stress-related illnesses, we would be wise to take notice of. But this is only the beginning of how a relationship with the natural world affects stress.

THE RECOGNITION THAT STRESS can make us ill is a relatively new aspect of Western medicine. The word "stress" was only introduced into medicine in the 1920s, by Walter Cannon, the American physiologist who first described the fight-or-flight response. It is thought that our physiological system first evolved to respond rapidly to a physical threat, such as the sudden presence of a wild animal, but is now activated much more regularly and consistently. The bills need to be paid, the emails need to be answered, the packed train is delayed, the fears of terrorist attacks and car accidents need to be managed, people are trolling you on social media—modern life is a cacophony of adrenaline-raising situations that can trigger the sympathetic nervous system.

The problem today is that we have fewer opportunities to reset the nervous system and to lower the level of cortisol, and thus there are many people in the industrialized world who suffer from chronic stress. If we are stressed, and on high alert, the nervous system is out of whack, which can lead to autoimmune disease, diabetes, cancer, cardiovascular diseases and mental illness—all the health problems that have overtaken infectious diseases as the leading causes of disability in the modern world. Increased autonomic nervous system arousal is also associated with sleep interference, which affects immune function and the overall health of the human body and mind. What can we do to balance the nervous system and soothe the stresses of modern life?

In the early and mid twentieth century, respectively, two groups of Britons who experienced extreme stress, and often sought succour or relief in nature, were the soldiers and internees of the First and Second World Wars. Companion animals were looked after, for therapeutic reasons, including cats, ducks, goats and rabbits, guinea pigs, working dogs and horses. More exotic pets included a fox, a baboon and a golden eagle. Canaries were frequently given to soldiers recovering from injuries. In wartime poetry and diaries, one can see the psychological benefits of nature to these men, many of whom had never been immersed in nature to the extent they were on the Somme, for example. Key characters in the writing and poetry from these wars were often birds—lapwings, rooks—as well as the trees and the forests. Birds, and birdsong in particular, were uplifting for those in combat; they were often described, in therapeutic terms, as solace for the soul. In his poem "February Afternoon," birds symbolize constancy for Edward Thomas, a reassuring sign that nature is still the same now as it was a thousand years ago: ". . . when one, like me, dreamed how / A thousand years might dust lie on his brow / Yet thus would birds do between hedge and shaw."

Soldiers also turned to gardening in the trenches. In the First World War, men grew flowers in the thick, damp soil on the front line or in makeshift pots made out of spent German howitzer shells. Forget-me-nots, cornflowers, sweet william, poppies, primroses and other wildflowers were planted from seeds collected in French villages or sent over by relatives.

At the start of the First World War, around five thousand mostly British civilian men were trapped in Ruhleben internment camp, near Berlin in Germany, a bleak, cramped and melancholy old racetrack with no beds or comforts. They wanted to garden, and fought and begged to be allowed to bring nature into the camp. Michael Pease, a keen botanist and geneticist who campaigned for the garden while he was imprisoned, led the calls. "No self-respecting Town Council in England is without its Parks and Gardens Committee; surely in Ruhleben, too, some public energy, and public funds, could be devoted to beautifying the Camp?" he wrote. He argued that bright colours "would be a source of untold joy to all throughout the summer and autumn." The civilians were granted their wish and grew a vast range of flowers, from asters to dahlias, nasturtiums to petunias and begonias to sweet peas, with seeds, bulbs and instructions sent in from the Royal Horticultural Society offices in London. Flowers weren't, as one might imagine, a luxury for internees; they symbolized an act of psychological resistance and much-needed hope. Later, a significant amount of food was grown—thousands upon thousands of tomatoes, lettuces and leeks.

Why did both the soldiers and the internees bother? The beauty of flowers must have been a distraction and a joy—they may have reminded the men of their gardens at home, their families and loved ones, as well as of Britain and its green and pleasant land. Planting might have given the men a sense of hope for the future at a time of extreme uncertainty. To plant seeds is to believe you will see them grow. The leaves would uncurl, the birds would sing, the flowers would open; nature would endure long after the war had ended. They would even have unknowingly benefited from those antidepressant properties of soil that O'Brien, Rook and Lowry discovered decades later.

When the internees and soldiers returned home, many continued to seek solace in nature. Gardens were given priority by local authorities in charge of building new housing after the war. In the United States, gardening programmes were also offered to military veterans with shellshock, or what we would now call post-traumatic stress disorder (PTSD). As they had done in the trenches, soldiers rebuilt and soothed their minds by watching plants grow and appreciating their beauty.

Today, there are horticultural programmes and outdoor initiatives for veterans and ex-service people to aid rehabilitation from injury and

PTSD all over the United States and Europe. Hiking in the wilderness is currently receiving particular attention to see if it could be effective in helping to treat PTSD. A cross-discipline research team at the College of the Environment, University of Washington, has paired with the Sierra Club's Military Outdoors programme to run two pilot studies, with plans for a clinical trial in 2020. Veterans with PTSD who join surfing programmes also report improvement in their symptoms. In San Diego, the navy is spending $1 million to research the effects of surfing on PTSD at the San Diego Medical Center.

To understand why people recovering from extremely stressful situations might turn to horticulture or hiking for therapeutic reasons, or why a relationship with nature can help with daily stress, we need to understand how spending time in nature affects the nervous system.

THERE IS A BENCH on my favourite local walk along an abandoned canal. It is set back from the path and flanked by spindle trees. In winter, on the ground around it, there are snowdrops; in early spring, the ramsons appear, filling the air with the smell of wild garlic. The view changes throughout the year, but one cold, grey wintry day I sat there and stared in a calm daze for a while. Robins sang; the breeze gently moved the catkins. Across the way, a small deer nosed through the undergrowth. Ducks swam quietly. I felt tranquil and serene and, afterwards, rested.

We are starting to understand how this type of activity can directly affect the nervous system and thus the way we feel. When we walk in the woods, or by a lake, or spend time in a garden or park, evidence suggests that our parasympathetic nervous system is more likely to be activated. After exposure to nature, our stress recovery response is faster and more complete when compared with exposure to built environments. This has important consequences for our health at a time when stress-related diseases are on the increase.

How, exactly, does it work? The parasympathetic nervous system slows the heart, dilates blood vessels, increases digestive juices, constricts the pupils and helps you feel calm. It is the "rest and digest" processes at work inside your body, associated with feelings of contentment, sleep and safety. High resting levels of parasympathetic activity have been found to have many benefits to our health, from emotional regulation to decreased risk of cardiovascular disease. The sympathetic nervous system's main function is to stimulate the body's reaction to stress ("fight-or-flight" mode) and is associated with vitality, appetite, excitement and drive. When the sympathetic nervous system is activated, the body ignores any non-essential business, such as immune function.

Ideally, then, we want a balanced nervous system. An unbalanced system—when the stress response system is activated too much, or left on for too long—leads to physical and mental health problems. One of the mechanisms by which nature makes us mentally well is its effect on our parasympathetic nervous system. It suggests that if we, as a society, are allowing trees to be cut down, or natural spaces in urban areas to be paved over, we are acting in a way that is damaging to public health. We really need nature in order to recover from the stresses of life.

A meta-analysis and systematic review conducted by Miles Richardson and Kirsten McEwan of the University of Derby's Nature Connectedness Research Group, published in the journal *Evolutionary Psychological Science,* found that when people were in a natural setting compared with an urban area, they had greater parasympathetic nerve activity and lower sympathetic nerve activity. They were more likely to feel soothed, content or calm. There are many studies that link nature sounds—and particularly a diversity and richness of bird sounds—to decreased stress and a quicker recovery of a balanced nervous system. Even people under anaesthetic have been found to produce fewer chemical biomarkers associated with stress—such as amylase in saliva—when played a recording of soft wind or birdsong.

But the opportunities to hear the sounds of the non-human world are reducing. Bernie Krause, an American musician and soundscape ecologist who has recorded the natural world since the 1960s, said that in 1968 it would take him ten hours of recording to gather one perfect hour of nature's sounds without any human noise that was good enough for an album or film soundtrack. By 2001, it took him up to a thousand hours or more of recording. "Fully 50 percent of my archive comes from habitats so radically altered that they're either altogether silent or can no longer be heard in any of their original form."

Other studies investigate the impact of the scents of nature on the nervous system. A Japanese study found that smelling cedar wood was associated with parasympathetic nervous activity and decreased heart rate and thus a state of physiological relaxation.

WE ARE LOSING the benefit of natural sounds, then, and natural smells and also natural light, which has serious consequences for psychological health. The rotation of Planet Earth is knitted into our bodies, through years of evolving outdoors, as if we had tiny watches in each of our cells. Synchronized with solar time, our internal clock allows us to anticipate sunrise and sunset. It doesn't just help us get a good night's sleep and wake up, it also conducts various elements of our biology, physiology and behaviour, from metabolism to hormones, from blood pressure to cell repair; telling us when to rest and when to use energy, and when to replicate DNA, the building blocks of our body. It makes the Swiss Vacheron Constantin Reference 57260, the most complex watch in the world, look like a clock made of Plasticine.

Research into the effect of circadian disruption on health through lifestyle or working patterns suggests it is not good for us at all. Studies of night-shift workers in Denmark found a 40 percent increased risk of breast cancer. It is thought that the hormone melatonin, which is suppressed by light at night, is involved in preventing damage to cells.

There is now a growing body of research linking circadian rhythms to mental health. Disruption of daily rhythms is strongly associated with mental health problems, low levels of happiness and the likelihood of major depressive and bipolar disorders. It isn't yet clear how or why the disruption affects our mind, though one possibility could be a lack of vitamin D. A literature review published in 2017 found a significant relationship between vitamin D deficiency and depression—and insufficient exposure to sunlight is a risk factor in vitamin D deficiency.

To look at working cultures in the West, you'd think we believed that we could somehow override our circadian clocks, or assuage any unhealthy consequences of disruption. If I'd said to my boss when I worked twelve-hour days in an office with very little natural light, "I can't work night shifts because I need to protect my internal clock by spending time in natural light in the day and avoiding melatonin-suppressing blue light at night," I'd have been handed my pink slip.

·   ·   ·

WHEN I FEEL STRESSED-OUT, or flat, I go for a swim in a river. I love the damselflies with their turquoise prom dresses glowing in the shade, the sunlight leaking through the swaying sycamores and the possibility of the neon jag of a kingfisher zipping overhead. If it is May, I might find myself swimming with mayflies on their one day of life on Earth, pirouetting up and down on the surface—tiny, pearlescent wraiths. The cornflower blue of the sky and the green of the poplars are pleasing to look at while floating along with the current. I love the smell of water mint. And of course, a cold sharp shock also jump-starts my system.

Wild swimmers often report improvements in depressive symptoms and an increase in the "animation and vigour" that diarist Fanny Burney attributed to early morning swims in the sea in 1782. Recent studies have linked cold-water swimming to a decrease in tension and fatigue as well as improved mood. Cold-water swimming acts as a stressor on the body, which activates the sympathetic nervous system and increases hormones such as noradrenaline. Possibly, this explains the feeling of endorphins which wild swimmers chase.

But there is another reason—and it's relevant even if you're not one for swimming in cold rivers: there may be something not just in the water, but in the air. Negative ions are more abundant around water and natural areas—rivers, beaches, waterfalls, mountains, anywhere that air molecules are broken apart by moving water, crashing waves, moving air, sunlight or radiation—than in indoor, air-polluted, air-conditioned areas, where the number of negative ions falls significantly. Humans ingest negative ions through their airways. Crucially, negative ions have been linked with biochemical changes in humans that activate brain activity and release serotonin, improving mood disorders and having antidepressant effects. Like the smell of cedar wood, they are said to activate the parasympathetic nervous system, which calms down the body and the mind. It could be one of the reasons surf therapy is growing in popularity.

THE VITAL LINK BETWEEN the nervous system, immune function and mood is becoming increasingly clear. Recent advances in genetics are starting to show that genes related to depression are also linked to the nervous and immune systems. Many benefits of connecting with nature enhance immune function, from relaxing the nervous system to the reduction of worry or anxious thoughts; from providing a break from the effects of air pollution to the presence of phytoncides (the chemicals emitted by trees and plants, which may boost the immune system). Studies have shown that just looking at a natural scene can boost levels of anti-inflammatory cytokines. Spending time in nature can increase resilience to stress and thus reduce the risk of inflammation when it isn't required.

A leading expert in the field of nature and health, Frances "Ming" Kuo, argues that enhanced immune function is the central pathway to explain the benefits of nature contact on mental and physical health, aided by the effect on the parasympathetic nervous system. "When we feel completely safe, our body devotes resources to long-term investments that lead to good health outcomes—growing, reproducing, and building the immune system . . . When we are in nature in that relaxed state, and our body knows that it's safe, it invests resources toward the immune system."

The most interesting and exciting research into this link has been conducted around the Japanese practice of *shinrin-yoku*. In the early 1980s, the Japanese Forestry Agency made a formal proposal that citizens should take mindful, multi-sensory walks in the woods to improve their health. It was named *shinrin-yoku* in January 1982 by Tomohide Akiyama, the Director General of the Forestry Agency, and made public in newspapers soon after. The phrase translates as "forest bathing" and the practice is rooted in ancient Shinto and Buddhist ideals of harmonic balance with nature, and engaging the five senses. Inhale the scent of the pine, the maple and the cypress. Listen out for the pipe of the cuckoo, or the rustling of a black bear. Touch the fuzzy bells of the Japanese mugwort or the bark of a tree. See the gingko leaves deepening to a buttercup yellow as the summer fades. Taste the fresh geosmin on your tongue.

In 2006, Iiyama Forest was given the first forest-bathing certification, and since then sixty-two forests have been certified in Japan. The practice is embedded in Japanese healthcare and is a normal part of the way people manage their health and stress: doctors will prescribe "forest medicine." South Korea also promotes forest therapy, and the government recently invested $14 million in a National Centre for Forest Therapy, where it is training five hundred forest-therapy instructors.

Researchers are continuing to explore why people report lower levels of stress and higher feelings of positivity and well-being after a walk in a forest or woodland. A forest therapy group founded in Japan in the 1980s began to investigate how forest bathing reduced stress levels considerably. Qing Li, from the Nippon Medical School, the leading scientist in this area, found that the forest boosted the immune system by increasing the number of natural killer (NK) cells and intercellular levels of anti-cancer proteins, possibly through phytoncides, the chemicals emitted by the trees and plants in the forest. In Asia, the Hinoki cypress and the Japanese cedar emit phytoncides, but there are many trees in Europe that do too: oaks, beech, birch and hazel, for example. NK cells are white blood cells that can attack and destroy cells that might cause damage; people with a lower risk of cancer and other diseases tend to have higher NK activity. Li also found that increased NK-cell activity lasted for a month after a walk in a forest, but not after a walk in the city. Other psychological benefits included decreased levels of anxiety, depression, fatigue and confusion, and increased vigour and energy. "If you want to decrease your stress, you have to go to the forest," said Li.

According to Li, forest bathing can be done anywhere there are trees, even a small garden, but you should be there for at least two hours. Deep breathing is important to ingest the forest molecules which increase activity of NK cells. The sheer scope and range of benefits Li and others have discovered makes it increasingly urgent to halt the destruction of woodlands and forests, for the sake of the health of future generations.

But, despite the evidence I've presented so far, how many of us would choose to go for a walk in the woods when we are depressed or anxious? It is all very well reading that walking in a forest for a couple of hours is good for your mental health, but how easy is it for most people to access actual forests? As the field of formal nature therapy develops, should contact with the wild be seriously considered as a drug-free alternative treatment for various mental illnesses?

# 5

# Plant Wisdom

I saw it as a seed, it's a journey, one point to another point, hopefully coming through. If seeds can do that, hopefully I can too.
—Patient in an NHS secure unit, 2018

I go to
nature because man is scary
—A. R. AMMONS, "The Ridge Farm"

ALL OF US have bad days and good days, and all of us are susceptible to stress, depression and other mental health problems. The evidence I have noted so far shows the degree to which we have overlooked how healing and uplifting nature can be for many people through various pathways—from the immune system to the cerebral cortex, and from the nervous system to the gut–brain axis. But what about people who are in the midst of a mental health crisis, or battling with chronic mental illness? Can nature work as a supplementary treatment? If you are suffering with highly critical or paranoid thoughts, or panic attacks, for example, can kneeling down to examine the first croziers of spring ferns, or looking up to take in a bright mass of blossom, relieve emotional pain, or make it easier to bear?

Ecotherapy projects are currently on the rise in the West, from wilderness therapy and pilgrimage walks to woodland therapy sessions and gardening groups. The term "ecotherapy" was first used by Howard Clinebell, a Methodist minister and professor of pastoral psychology, in the early 1990s, as part of an effort to incorporate an ecological component into his counselling work. He argued that "ecoalienation" led to alienation from "our minds, souls and relationships." Being "nurtured by nature," he wrote, "helps heal the agonies of grief in general." In various books, he set out practical ways of integrating nature into therapeutic settings. The concept spread rapidly through the 2000s, and even more widely in the 2010s.

In Britain, in both the NHS and the private sector, there is a drive to offer people with mental health problems community activities that involve being outdoors, as the evidence for the scientific benefits of a relationship with the natural world accumulates. Mind, the mental health charity, funded 130 ecotherapy projects in England between 2009 and 2013 with a £7.4 million grant from the Big Lottery Fund. "People who have attended ecotherapy programmes have reported that it has reduced their stress levels, reduced their depression and improved their self-esteem," according to the Mind report. But despite early positive reports, such as a Swedish paper which found that nature-based rehabilitation programmes led to a reduction in healthcare consumption, the number of green-care social prescriptions in Britain is currently low. The Clinical Commissioning Groups that run the NHS budget rarely fund these projects, so they must find money independently—an obstacle to scaling up green care for a greater number of people.

And yet, even so, a new movement advocating the prescribing of nature-based interventions for health is growing. Green gyms, walking-for-health programmes, care farms, forest schools, green-therapy groups, surf therapy and horticulture initiatives are spreading. In 2017, an accord between National Parks England and Public Health England was set up to represent a commitment to improve health and well-being through use of national parks. Projects include an outdoor activity group in the Peak District for young people who have had their first episode of psychotic illness; walks on the South Downs for people with mental health problems; and recreational events on the Yorkshire Moors to lift spirits and encourage activity. Mind is also training park staff in how to work with people with mental health needs.

Some British academics I spoke to expressed concern that the social prescription movement was moving too fast without adequate evidence of its benefits. "The evidence base for what nature-based interventions are doing when they're out there is really fragmented at the moment," said

Dr. Rebecca Lovell, Research Fellow at the European Centre for Environment & Human Health, University of Exeter. "Plausibly they're very beneficial. They incorporate activity, social contact, getting people out of their normal potentially destructive context. But we don't know really if this delivery mechanism is a good way of using the environment to promote people's health."

The American physician Howard Frumkin was more sanguine. "When it comes to nature contact, if there were a false positive, what's the harm? People get outside, they may exercise, they may feel restored. Nature contact isn't completely risk-free; think of sunburn, bee stings and allergic reactions to pollen. But across the population these are relatively minor. In medicine in the last generation we've become accustomed to being very demanding about the level of proof that we need before advocating and implementing an intervention. That makes perfectly good sense with medications and surgeries, which carry large potential risks, but we can be more forgiving when it comes to nature contact . . . Public health history is full of examples of understanding that something has a beneficial or dangerous effect without clarity about the pathway, but with enough certainty that it's time to act," he said.

Concurrently, what we do in natural spaces is changing. While woods were once used for fuel, timber and hunting, now you might find community healthcare projects, forest schools, art projects and conservation projects with a therapeutic element. Since 2010, Gendered Intelligence, a Community Interest Company, has run summer camps for young trans people in England, a group that is at a high risk of mental health problems and suicide.

While GPs in the United Kingdom can write "social prescriptions" for people with a range of physical and mental health problems—from mild to moderate to long term—the activity itself is carried out by the voluntary or community sector. What a GP can prescribe will differ from region to region. In Cumbria, for example, there's a working farm which takes referrals from any adult with an existing mental health problem. In Cornwall, sailing or surfing projects specifically designed for psychological support are offered. One of the most common forms of nature-based interventions is horticultural therapy. Since the 1980s, the number of initiatives that offer gardening to help people with learning disabilities and mental health problems has grown significantly. Of course, the idea is nothing new: Florence Nightingale herself, in her notes about flowers, was drawing on a rich history of perceiving gardens as therapeutic. But since the 1980s the idea has been formalized, with training, certification and evidence-based research backing up the results of horticultural therapy. One of the pioneers and largest projects is called Thrive and its headquarters is near Reading in Berkshire, just up the road from where I live.

IN THE 1970S, a British man called Chris Underhill was working in Zambia with Voluntary Service Overseas. He noticed that contact with nature—plants and the land—helped the people he was working with who had learning disabilities and mental health problems. When he returned home to Britain, he set up the Society for Horticultural Therapy and Rural Training in Frome, Somerset. To begin with, he visited hospitals, secure units and day centres in order to adduce horticultural therapy to the mental health services. "Some of these rather large mental hospitals, as they were then known, all came with lots of land that people didn't use, or rather were unclear how to use in order to help their patients . . . I wanted to introduce a new way of gardening and horticulture with a more disciplined approach and a will to do things properly—like wearing the correct clothes and shoes, using the correct tools and creating an understanding about what they grow they will harvest and then cook."

The Society set up its headquarters in a magnificent Berkshire estate, bequeathed by a family friend of Underhill's, a parish priest who had witnessed the powerful effects of gardening on his parishioners. It was renamed Thrive, and it's now the leading gardening-for-health and horticultural therapy charity in the United Kingdom.

Thrive sees an average of eighty clients a week with various disabilities, such as learning difficulties, mental health problems, PTSD, autism, dementia or visual impairment. It has helped thousands of people and was one of many similar projects across the country making a measurable difference to the lives of people with mental health problems. In 2017, it directly helped 1,100 people at the Beech Hill National Office and outreach projects in schools, care homes and hospitals. Most clients work at Thrive for one day a week, but for some it's up to four days. A person with profound multiple learning difficulties might stay for a while;

someone with a mental illness might come for a recovery period before returning to their job. There are seven therapists, who come from different backgrounds in social care, teaching, horticulture or other occupational therapies.

The aim of the horticultural therapists at Thrive is to use plants and the gardens to improve the physical and mental health of people with a disability or illness. How? First, by bringing people out of their homes to connect with other people, which is often one of the hardest things to do when your head is spinning and the urge to isolate yourself is strong. For some of the people who attend Thrive and other, similar nature-based interventional projects, it is their only contact with the outside world. Second, by giving people purpose and achievement, through seeing and watching their flowers and plants grow and produce fruit, which can be turned into soups or crumbles. Thrive claims that this type of therapy can help people recover from all sorts of mental health conditions and also "slow down deterioration of a degenerative illness."

It is autumn when I visit, and everything is on the turn. The three acres of gardens are divided into smaller areas to make the work more manageable and to enable horticultural diversity. The green and white tranquil area is a place for people to spend time if they want to avoid sensory overload. The potager garden has a mixture of ornamental plants and edibles, such as sculptural sweet corn and cavolo nero, which explodes ostentatiously out of a bed. Everything is maintained by the clients (the term Thrive uses), who also have their own personal beds, and they can take the harvest home. One designed the plot with the colours of his favourite football team. In the "journey garden," the planting is lighter, with black bamboo and Black Lace (*Sambucus nigra*) at the entrance. A garden for the visually impaired features grass that makes a whirring sound as it moves in the wind, stark and contrasting colours and the soft-to-touch lady's mantles. The Japanese garden has a large pond. Every area is its own magic kingdom of shape, colour and texture. It is deeply thought-through, compassionate and moving to behold.

"Run your hand through and smell that," said senior horticultural therapist Jan Broady to me, pointing to Salvia "Hot Lips." He is a smiley, passionate ex-teacher with a soft tone of voice. "It's a real blackcurrant smell." He was right: the petals puckered like the lips of a showgirl and it smelled like minty Ribena. Broady and the other horticultural

therapists don't engage in conventional talk therapy while gardening—
the gardens are their primary tool and they are not trained or accredited
psychotherapists—but, as he put it, they are always asking how people are
feeling and ensuring that they feel heard. If a client did need a trained
psychotherapist to speak with, they'd recommend another service.


ONE SYMPTOM OF DEPRESSION IS anhedonia, the inability to find
pleasure in normally pleasurable activities. I had experienced dark peri-
ods even after I rekindled my relationship with the natural world, when
I barely responded to birds or trees or flowers. A few months after we
moved to the house in town with the garden, I was diagnosed with post-
natal depression and prescribed an antidepressant SSRI. Until my symptoms started to improve, I lost any interest, joy or delight in nature. I had
very little response to trees, clouds or the colours of the leaves or birds. I
felt nothing in wild places that used to fill me with a deep peace. I continued to go to the wellspring, walking in a daze with my baby in the
pram, and hoping that my emotional reaction would be restored, but it
was chilling to lose that connection for a while.

One winter morning, sleep-deprived, anxious and psychologically
strung out, I walked out into the garden, baby bundled up warm in a
suit and hat, and into the path of a deep, heady, orangey, sugary perfume.
I couldn't believe there could be a smell so strong at this time of year. I
followed the molecules in the air with my nose and noticed a bush with
compact pink flowers. I pulled one off to smell it up close. It was a bright,
clean, Cleopatra-strong scent. I put it to my daughter's tiny acorn nose
and I'm sure her eyes widened slightly. I looked up and a red kite circled
above, its russet finery blazing against the blue. I fetched scissors from
the kitchen, cut off some of the branches and stuck them in a pot in the
kitchen. With a little online research, I discovered that it was a Daphne
bush. Each time I smelled the flowers, I felt hopeful, reassured, connected.

A couple of months later I was in Scotland at my mother's. She lives
near the Galloway Forest Dark Sky Park, under some of the darkest skies
in Britain. I was finding sleep-deprivation and the psychological transition into motherhood tough. At around midnight, after hours of rock-
ing indoors, I decided to walk my baby out into a small woodland. The
shrubs and bushes were crackling with . . . droplets? Insects? I heard a

larger creature snuffle in the undergrowth. But the real attraction was above me. Light-years and seams and skirts of stars in varying levels of brightness and size and twinkle. The sky seemed blitzed with silver. I couldn't tear my eyes away. I searched for the Milky Way and saw a shooting star and felt part of something fixed and bigger.

To cultivate enthusiasm in depressed client gardeners who are struggling with a loss of joy, Thrive makes sure they have their own area to plant, which can be motivating, but also means if they prefer to be alone on their plot, they can be. Talking and interacting for many people with mental health problems, especially introverts, can be draining. Nature, though, requires and demands nothing of this kind.

The horticultural therapist would start with a bit of direction, suggesting seeds which might give a speedy result. They'd also chat with the client gardener and come up with an individual development plan, based on their favourite colour or food to eat, or something they might relate to or look forward to. If plants failed, as some do, this could be an instance of learning that, as Broady put it, "things die back, but we will see them again. It's learning to cope in a small way."

There were hidden benefits to gardening or working outside, said Broady. One was simple: being outside helps people sleep better by regulating their circadian rhythms, and sleep is an important factor in poor mental health. Smell, too, is a powerful trigger of memory, so people with dementia would often work with lavender, because of its strong scent. "One of the lads was starting to say, I hate lavender, I hate lavender," remembered Broady. "It turned out he'd been working in a polish factory and it wasn't a happy time. We suggested he go and do something else, but he wanted to stay and said it was good for him to be remembering."

Horticultural therapists are given specific prompts to use that might encourage reminiscence among people with dementia. These include childhood games, such as making daisy chains or making a wish while blowing on a dandelion clock; folklore and Cockney rhyming slang ("daisy roots" for "heavy work boots"); old remedies, such as dock leaves on nettle stings; and local names for plants and flowers (cowslip, for example, is also known as "Our Lady's bunch of keys" or "galligaskins").

Some in horticultural therapy benefit from a non-judgemental relationship with other living things at a time when they might be struggling with personal relationships. They take a seed, water it well, give it some food if it needs it, plant it out with care not to damage the roots, in the right kind of environment, and so on—and the plant thrives and survives. And if a plant dies, this isn't a moral judgement, or a reward or a punishment; it's simply a consequence of the natural order of things. Gardening can also give a client a sense of agency, confidence and individual responsibility. This non-threatening but interpersonal element of the natural world was something that came up in many of my conversations. "I don't feel judged out here," said one friend, while we were on a walk. Nature doesn't do the modern currency of likeability. It doesn't allow for "likes" or retweets or favourites.

Andrew, a tall, middle-aged man, had been at Thrive for a couple of years. He was busy making bunches of roses, his favourite flower. How did the work make him feel? Happy, he said. "Getting outdoors. I don't want to stay indoors all the time. Very happy, meeting more friends." He beamed, and looked relaxed and serene.

I felt the same while I was at Thrive: charmed by the beauty, delighted by the taste of the reddest apple with the whitest flesh picked fresh from a tree, excited by stories of recovery and wellness in this rich, vibrant garden.

Why and how do the depressed at Thrive start to feel something when they see the seeds grow? The will of nature, its strangeness and its abundance, its drama and adventure, is potent. Nature is busy. It renews and it fights. It wants to live. Fungi grow and grow and grow like the ears and noses of old men and women. The emerald cockroach wasp (*Ampulex compressa*), a beautiful green and crimson parasite, injects an elixir of chemicals into a cockroach, which, zombified, is led, like a dog on a lead, to the wasp's nest, where the wasp lays its eggs in the cockroach's abdomen. So tenacious, so fervent, so driven. Isn't this why we love snowdrops? Why collectors will flock to gardens and woodlands to search for the rarest kinds? The yellow-marked that gleam in the sun; the green-splashed; the green-dashed on a frilled segment? Tiny variations that can persuade people to part with hundreds of pounds—up to £1,350—for one bulb. In the depths of winter, after Christmas cheer has been swept away, in those dark, ascetic months of January and February, snowdrops push through the ground and give us hope that spring is close. But not only that; snowdrops show the persistence and doggedness of the natural world. When we feel tired and run down by the short days, for those of us who live in the Northern Hemisphere, the snowdrop exhibits a determination and spirit that is encouraging, I find, even on an unconscious level. I stand and stare at the carpet of snowdrops in the wild cemetery next door and feel both a peaceful embrace and the "force that through the green fuse drives the flower," as Dylan Thomas put it; that we are getting closer to the sun and all her attendant abundant chain reactions.

I think often of Mrs. Alice Blunden, who was buried alive twice— twice!—in my local cemetery in 1674. The town was fined for negligence. When they disinterred her after the first burial, her nails and limbs were broken and bloody from trying to escape. Tenacity, it is in us all.

TENACITY IS JUST ONE of many useful qualities that can be found in the living world. At mental health and ecotherapy groups I visited while writing this book, I heard about a wide and creative array of images and attributes that people had discovered in nature and been able to utilize in their own lives. One woman had struggled with a difficult relationship with her brother for decades. She started to think about plants—"plant wisdom," she called it—and how, unlike humans and other animals, plants can't change their environment and move away. They have to adapt to their surroundings. If it rains, they can't go inside, so they have to survive. She was inspired by the plants to find a way of adapting to the situation with her brother, and one day she woke up to find her attitude had changed. Another person took daily botanical walks to distract himself from what was going on in his head. He would diligently notice and identify the plants, and how they changed, grew and adapted over time. He drew his own strength and resilience from the verve that drove the plant to adapt, and found the patience and slowness required in his walks a calming antidote to the rush of the modern world.

"Ecopsychology" is used by different people to mean different things. American counter-culture historian Theodore Roszak, who popular- ized the term in his book *Ecopsychology: Restoring the Earth, Healing the Mind* (1995), defined it as "the study of how ecology interacts with the human psyche." It's associated with deep ecology, Gaia psychol- ogy or Earth-centred therapy. The destruction of the rest of nature is a key facet of ecopsychological thought. However, some people use the term interchangeably with "ecotherapy" to mean taking the therapeutic relationship—usually one-on-one, but also group therapy—into the natural world.

Ecopsychology is still on the fringe compared to other psychological disciplines, such as psychodynamic psychotherapy, psychoanalysis or cognitive behavioural therapy (CBT), but more and more therapists are taking their work outdoors. Plainly, it is a mistake for us all to ignore the fact that we live within a wider ecology, that we are interconnected, and often engage in a poor or abusive relationship with other life forms, but you can see why ecotherapy is not mainstream: Western psychologists have barely given the relationship much thought (apart from Carl Jung, who we will come to later). We are indoctrinated into a life without nature from a young age, and for those of us who don't like being outside, a therapy session outdoors seems unlikely to work.

But ecopsychologists I spoke to extolled the benefits of being beside a river or stream, for example, rather than in a traditional consulting room, for getting some patients to relax and focus. My only experience of the therapeutic process is face-to-face and inside, but I often find it easier to listen or open up to close friends and family while walking somewhere beautiful under open skies, sharing and bonding over the weather, or what's come out that season, or the smells and sensations around us. I can often say things better beside a tree than behind a closed door.

John Scull, a retired ecopsychologist based on Vancouver Island, spent the last six years of his practice taking patients, many of whom were suffering from chronic pain and anxiety, into the natural world. He discovered that nature was a better healer than he could ever be. "People would come into the office and they're all completely wrapped up in themselves. What you're doing out in nature is helping them wrap into something larger and more extensive," he said. On a practical level, he would send people out to look for objects that suggested interconnections, or change. When they returned with what they'd found, they'd discuss how it might relate to their life and specific situations. For example, an older woman was depressed about ageing and in recovery from cancer treatment. The walk led her to a group of trees. "The older woman was amazed at the idea that when we look at trees, we value old trees much more than we value young trees. For her, that was an amazing insight about her life; no one bothers about ripping up a little sapling, but you're not going to rip up a six-hundred-year-old fir tree. For her—and for me for that matter—it was a major insight into valuing age differently."

Don't we also see shared qualities in the natural world, if we look closely? To stand with an ancient oak is a lesson in patience, in the slow-burn, in trust, in taking things one day at a time. It is to be in the presence of eternity and infinity, a universe much larger and cleverer and more complex and crazy than the eye can see or the brain could imagine. Aspects of the natural world also change, morph and have renewed significance in line with the unfolding narrative of a person's life. What gives a person resonance as a new mother, as a divorced person, as a young adolescent, as a person at the end of life, will be different, but there is an abundance of parallels and stories out there in which to find ourselves.

AT THE BEGINNING of my journey, I'd presumed that nature as a therapy, via gardening or woodland work, or just generally as a salve, was a slightly soft alternative treatment for the "worried well," or those suffering from mild to moderate levels of depression and anxiety. To find out whether it could possibly touch the sides of the more severe illnesses and conditions such as psychosis or schizophrenia, I visited an NHS secure unit for people with severe and complex mental illness.

The entrance was a glory of blue and yellow stained glass. The reception area furniture was constructed in wood from the trees that had stood there before the clinic was built. I was greeted by Sarah, a Horticulture Instructor whose work is part of the occupational therapy provision in the unit. Her eyes were clear and wide, and she smiled with her whole face. She was warm and likeable. She led me through the security area—visitors are not permitted to take phones or recording devices into the unit—and my bag was placed in a locker.

Sarah's sessions are offered to service users from two separate units. She works both in a specially designed occupational therapy garden and on the wards of the two clinics. Her aim is to use gardening as an opportunity to promote well-being. One key aspect of this is to teach people to grow and prepare food from "plot to plate." "When people's hands and minds are busy planting seeds, there is a focused engagement, a kind of release from pressure," she says. "It is possible to have a different kind of conversation."

Most of the service user group in the units have experience of the criminal justice system. The purpose of the units is both to manage risk and to support the recovery of service users with serious mental illness, who have often come from challenging backgrounds. At times this limits the choices that are available to people in their daily lives. Service users may stay for six months to two years, but some people are there for much longer.

Inside, the unit is dotted with artworks and sculpture, but it clearly remains a clinical environment. Because of the need to manage risk, both to themselves and others, supervision and assessment are a regular part of each service user's day. The more acute the illnesses treated in the ward, the higher the level of supervision and risk management.

The juxtaposition with the garden is clear, with its rainbow of colours, textures and smells. It is easy to imagine how being outside in the garden, with the hum of bees and the smell of flowers, the unpredictability of seeing a bird overheard, would be a sharp contrast to life inside the unit. "When you step out into this garden you understand the full sensory impact of being outside," says Sarah. "In this garden, in the here and now, all the processes of life are happening."

Asylums in the past, as well as mental health secure units, inpatient hospitals and prisons in Britain today, have a history of horticultural therapy. Since the First World War, "green care" has often been part of occupational therapy provision. Sarah sees twenty to twenty-five people per year, in groups of up to six people, often with additional staff support. There are strict rules about tool and sharps usage. She also takes gardening onto the acute wards, for people who are too unwell to access the clinic garden. The results have been positive.

Sarah, two service users and I sat in the sunshine, surrounded by flower beds and plants. We talked about their relationship to nature and gardening. Paul has always enjoyed gardening. He likes to produce his own food—broad beans, peas, onions—and has always found it therapeutic and relaxing. "It can make you feel calm and peaceful," he says. John sees gardening from a more philosophical point of view. "You plant things and see them grow and they never come the way you think. Nature is doing its own thing." He describes it as a primordial connection with the ground, touching the soil and, from that point, feeling grounded and down-to-earth. He talks about feeling joy in seeing his potager garden grow with potatoes, courgettes, pumpkins, sweet peas, broccoli and lettuce. "It gives you serenity, if you come from the ward where it's tense; you focus on the job at hand and become one with it; it makes you very serene."

Sarah explains to me the importance in this context of contributing by tending the plants and environment—which is the creative and responsible role of the "nurturer." Service users rarely have an opportunity to look after anything, but gardening offers them a chance to nurture, grow and care for a living thing. Sarah was witnessing something on a daily basis that the rest of us have forgotten, or are only beginning to comprehend: that we need nature on a psychic level—and without a connection to the rest of nature, the most unwell in our society will be even more disconnected. Any scepticism I had that a connection to nature was a silly suggestion as a treatment for people on the more severe spectrum was quashed.

Gardening gives service users a purpose, focus and a chance to become more resilient. "Sometimes nature doesn't do what you want; when things are difficult you adapt," says John. Sometimes the vegetables are not perfect, or seedlings fail or there might be an outbreak of pests, but the example of nature is helpful in teaching service users patience and resilience. After a gardening session finished, the feeling of peace would last for a while. John describes it as charging up the batteries to go back to the ward.

In a warm polytunnel, with sweet peas, pumpkins, coriander, tomatoes, strawberries, courgettes and fennel growing, Sarah tells me a story about a service user with severe psychosis on one of the acute wards. He was pacing around the room, and responding to unseen stimuli by shouting and singing. However, when Sarah put bedding for a plant container into his hands, he stopped, sat down and began to focus. He talked about his experiences with gardening and handled the flowers with care, choosing the colours, creating the design and then planting them. At the end of the session he had produced something with which he was pleased. "My sense is that if people are in secure environments without access to nature they become even more unwell," Sarah tells me. "I see the desire to grow things as intrinsic to being human." It reminded me of that statistic that many children spend less time outside than prisoners in solitary confinement. Patients at the unit are given greater opportunities to grow things than many schoolchildren have.

Why is it so healing? "Gardening is about relationships; it brings us a vivid awareness of a singular place in nature and how it changes with the seasons. It brings us into a relationship with what sustains us, with our own physicality, with our own story, with the people around us, and with a wider world," Sarah says. "To echo John, whatever our past or current challenges, it provides an interaction with the land, a lived experience of touching the earth. It is truly about reconnection."

The people I spoke with at Thrive and the unit, as well as other patients and ecotherapists elsewhere, convinced me that a relationship with the natural world must be built much more into our healthcare for mentally unwell people, and, as we will see in the next chapter, for the good of our wider society.

PART IV

# TRUNK

# 6

# Equigenesis

The weeds in a city lot convey the same lesson as the redwoods.
— ALDO LEOPOLD, *A Sand County Almanac* (1949)

From heavy hearts and doleful dumps, the garden chaseth quite.
— NICHOLAS GRIMALD, an Elizabethan poet, "The Garden"

ROBERT TAYLOR HOMES was a public housing project completed on Chicago's South Side in 1962. It was the largest development of its kind in the United States at the time. The estate was built on the plan of one building, which was replicated twenty-eight times across a three-mile stretch. To the west, there were railway tracks and an interstate highway; to the east, an eight-lane arterial road, so that residents were geographically quite isolated. At its peak, it housed 27,000 people, including the *A-Team* actor Mr. T, although the project was originally planned for 11,000 residents. The Mickey Cobras and Gangster Disciples gangs dominated the area and at one point sold around £35,000 worth of drugs every day. Violent crime was rampant too: a young woman was thrown from a window, an infant abducted and never seen again. "We are survivors. We have to be, given the dehumanizing injustices inflicted upon us each day," wrote resident Diana Robinson in a letter to the *Chicago Tribune* in 1987 about the experience of living at Robert Taylor.

From the photographs that remain, the Homes, with their uniform tower blocks, resembled "a gigantic vertical zoo, its hundreds of cages stacked above each other," as J. G. Ballard wrote in his fictional dystopia *High-Rise*. However, there was one difference. Some of the sixteen-storey buildings had trees and grass nearby or were near a park, while others were paved, barren "moonscapes" of asphalt and grey, "unrelieved deserts" of concrete, brick and glass. At first, trees and grass were planted in the surrounding environs of every block, but over time, to save maintenance costs, the local authority paved over the grass with asphalt or blacktop, and the trees eventually died in some areas. "Super hard, super hot and intensely urban," was how US landscape architect William Sullivan, who spent years researching the area, described it to me. Another called it a "no man's land."

In the late 1990s, the estate became the location for a number of studies conducted by the University of Illinois College of Agricultural, Consumer and Environmental Sciences (ACES) team led by Frances "Ming" Kuo (a cognitive psychologist) and Sullivan into the effects of physical environments on human well-being. The research continued until the estate was demolished in 2007, under the false promise of newly refurbished homes. The studies were early examples of empirical, controlled evidence that linked exposure to nature to human psychological health.

Kuo and Sullivan built on the early work of Bernadine Cimprich of the University of Michigan School of Nursing, who published studies in 1992 and 1993 measuring the impact of restorative environments on people diagnosed with breast cancer. Her studies were central in giving Sullivan, Kuo and other early researchers a sense that people living in difficult conditions might benefit from regular contact with nature or green spaces.

Sullivan already had a hunch that a connection to the natural world mattered. When he was growing up, his mother would say, "You've got housitosis! Get out of the house!" At the age of twelve, he got a weekend job with a landscape contractor. He started off mowing lawns, hauling bushes, raking leaves and cleaning up. As he got older, the responsibilities grew and he was struck by what a difference a greener, well-designed environment could make. "I thought, 'Wow, we're doing something really wonderful here. It's good for the environment, the people who work here and the people who pass by.'"

At Robert Taylor Homes, Kuo and Sullivan looked at whether trees had a measurable impact on people's lives. The typical resident—hundreds were interviewed—was a thirty-four-year-old African American single woman, unemployed, educated to high-school level, raising three children on an annual household income of less than $10,000. What they discovered was unequivocal: a relatively low dose of nature, a few trees and some grass close to an apartment, was enough to significantly improve the psychological well-being and cognitive functioning of their participants.

At first, they studied how groups behaved outside the development, to see if trees made a difference to social connection. They found that people were more likely to congregate around areas with trees than not. In this otherwise harsh urban setting, the trees created a comfortable environment for neighbours to talk, socialize and make friends. They also helped child development, for children were more likely to play and interact in areas with trees. "Imagine feeling irritated, impulsive, about ready to snap due to the difficulties of living in severe poverty," said Kuo at the time. "Having neighbours you can call on for support means you have an alternative way of dealing with your frustrations other than striking out against someone. Places with nature and trees may provide settings in which relationships grow stronger and violence is reduced."

As well as providing a venue for friendships to develop, Kuo and Sullivan believed that trees could also have a restorative effect on mental fatigue. Rachel and Stephen Kaplan, who developed Attention Restoration Theory (ART), were the dissertation chairs for Kuo and Sullivan. Kuo and Sullivan found that ART was the best explanation for the effect of the environment on the residents of the Robert Taylor Homes. Mental fatigue, caused by living in a hard, harsh environment, with few opportunities for restoration, could lead to irritability, which makes social relationships suffer. The energy and resilience required to cope with life in this demographic is high, they observed, and combined with the threat of danger and the demanding environmental conditions of noise, crowds and stimuli, a restorative time-out was rare.

"What we have here is a naturally occurring experiment which gives some confidence that the differences we found in social behaviours were due to the trees and not to other factors," Kuo told *The Illinois Steward* in 1999: apartments were assigned by chance; the scientists ensured that "better" residents had not been given "better" homes. To prove the link further, Kuo and Sullivan examined police reports for ninety-eight apartments in another large housing project in Chicago, the Ida B. Wells Homes, one of the poorest neighbourhoods in the United States at the time. They found that the greener a building's surroundings were, the fewer the total crimes. "Before we started our research, I would have said trees are nice, but the problems we're facing in our cities and our budgets are such that I'm not sure they're worth it," said Kuo. "I think that through this research, I have become convinced that trees are really an important part of a supportive, humane environment. Without vegetation, people are very different beings."

Following the research, Chicago's city government spent $10 million to plant twenty thousand trees in the city. It was a rare moment of civic investment in nature. Although we've known how essential trees are to health and well-being and functioning societies for decades, the recent destruction of green spaces and trees in places in the United Kingdom, such as Sheffield, and parks and other natural areas more generally, suggests we've either forgotten or don't quite believe it.

Speaking over a decade later on the impact of his work, Sullivan thinks that contact with nature is still regarded as an add-on. "It's nice if you can do it but it's not necessary is the misinformed view of too many people, and what the research evidence from now many hundreds of studies across continents continues to show us is the centrality of being exposed to green spaces within cities on a daily basis and the profoundly positive health consequences of that kind of exposure." The research at the Robert Taylor Homes adds clearly to the evidence: healthy, robust, well-functioning and happy societies require a connection with the natural world. It must no longer be seen as an optional frill.


IN 2018, a Philadelphia-based team ran a randomized controlled trial which looked at the effects of greening vacant lots both on levels of crime and violence, and on perceptions of fear and safety. About 15 percent of land in US urban areas is deemed vacant or abandoned, an area about the size of Switzerland. In Philadelphia rubbish and debris were cleared, trees were planted, grass seeds were sown and fences erected. The results of the intervention, gathered via police-reported data and anecdote, were a significant reduction in gun violence (29 percent), vandalism and burglaries in the areas. Residents felt less scared about leaving their houses to go outdoors and spent more time outside. Another study found that greening vacant lots reduced depression in residents of the city.

The presence of nature in prisons has also been found to reduce reoffending: gardening programmes in the United States have been found to decrease inmates' likelihood of committing crimes after release, and many prisons offer gardening projects and other nature-based activities. In the United Kingdom, the GOOP (Greener on the Outside for Prisons) project in the north-west of England reported better mental health for its prisoners, including increased confidence, self-esteem and self-control. In Halden Prison in Norway, biophilic design has been incorporated to improve the mental health of inmates and correctional officers, with natural light, wood furnishings and access to woodland.

Snake River Correctional Institution in eastern Oregon is about as far from snakes and rivers as you can get. Many—60 percent—of the three thousand inmates of the medium-secure prison spend time in solitary confinement. The time spent isn't just a few days here and there; it varies between seven months and three years. The cells are small (12 feet wide, 7 feet long and 8 feet high) and the prisoners stay in their rooms for twenty-three hours a day. Each cell has a bed, a toilet and a metal door with windows that face inside. It is the ultimate in sensory deprivation. Human rights groups condemn the punishment as an incubator for madness and, unsurprisingly, prisoners in solitary are more likely to kill themselves. It is estimated that eighty thousand people are confined in this way in the United States at any one time. At Snake River, inmates are allowed outside for forty-five minutes a day four or five times a week to exercise in a cement-floored yard. The yard is walled and enclosed: they can see the sky behind a metal grate but nothing else. They rarely see natural light, clouds, the sun, trees, leaves, plants, birds or insects. Prisoners in solitary confinement are among the most, if not the most, nature-deprived humans in the world. Many arrive in prison with severe mental health problems, which worsen while they are in solitary. Many try to take their own lives. Many lose their minds. It is hard to imagine a more dehumanizing environment for people who are severely mentally unwell, as over two-thirds are. Staff working in these facilities are also more prone to stress, anxiety and depression.

A correctional officer at Snake River watched a TED Talk online by a forest ecologist called Nalini Nadkarni, a biology professor at the University of Utah. She talked about her work in prisons teaching inmates how to grow moss and raise endangered frogs and butterflies. She found that the prisoners engaged with science and often found resonance in the natural world. "I taught the men that mosses are very resilient—they can dry down and stop photosynthesizing until they are rehydrated, even after years of no water," she wrote. "One of the inmates asked if he could take a sprig of moss to his cell, where he kept it in his drawer. Later, he said, 'I check it every day, and even though it lies in the dark, it is still alive. Like me.'"

The officer called Nadkarni up to see if there was anything she could do for the inmates at Snake River, who were difficult for staff to manage, with behavioural issues that required significant staff time and resources. She devised what she named the "Blue Room." She took over one of the indoor exercise cells, which was painted blue, and set up a projector. Inmates could drop in and choose various nature films featuring footage of underwater ocean scenes, forests, flowing rivers, tropical jungles, a burning fireplace, clouds or the Earth viewed from space. Some had classical music as a soundtrack; others had the sounds of nature. Her hypothesis? That simple images of nature—photographs or films—could have an effect on human well-being and mental health for the severely nature-deprived. It was called the Nature Imagery in Prisons Project and it began in April 2013. For the study, inmates were separated into two groups of twenty-four. One group was given the choice either to watch nature videos in the Blue Room or to exercise outside five times a week. The other group were allowed to exercise, but weren't given access to the videos.

In September 2017, she published her results. The inmates who had watched the nature videos reported feeling calmer and less irritable. They said the positive feeling lasted well after watching the video. Some reported feeling less tense; others said it improved their sleep. "It is temporary respite from a horrible environment," wrote one. Staff were initially sceptical about the intervention, but all respondents agreed that inmates calmed down after watching the nature videos. Those in the nature group had 25 percent fewer disciplinary reports for violent incidents, and staff observed fewer angry outbursts and incidents of self-inflicted injury. Reducing the number of violent incidents saved the prison thousands of dollars in medical bills. The Project is now being rolled out at prisons in Wisconsin and Nebraska.

Nadkarni and the other authors concluded that the study offered further proof of the need for a connection with nature for those who are extremely nature-deprived—from those in hospital settings, military barracks and space stations to office workers, road-toll workers and children with little access to the natural world. The vast majority of us, in fact.

HOW CLOSE WE LIVE to nature makes a measurable difference to our health. People who live near parks, forests or the sea report better mental and physical health. The incidence of depression and mental distress is lower and life satisfaction is higher when people live near nature, rather than in built-up urban environments. In studies, the elderly, housewives and people in lower socio-economic groups seemed to benefit the most from being near natural areas.

Nature is not a luxury: its presence or absence creates and causes different health outcomes for different groups of people. There is a direct benefit from being near nature to our mental health. Living among higher levels of greenery—trees and plants in urban areas—has even been associated with living longer.

So parks and green space are vital to human health and happiness. Fresh, natural, clean air should be a human right. Biodiversity, birdsong, forests, awesome natural phenomena, clean rivers . . . these aren't elements of Planet Earth we can do without if we want to live our lives to their full potential. But there is another problem, alongside ecological destruction. Within the winnowing landscapes and downward trends, there is a deep inequality in access and connection. And this is a stain on our society.

People don't suffer equally from environmental degradation, threat or disaster. Vulnerable and marginalized populations on lower incomes are often more exposed to pollution, hazardous waste and toxic chemicals. As the impacts of man-made climate change worsen over the coming decades, the poor, marginalized and vulnerable will continue to suffer the most. Children, women, the elderly and the sick will face the harshest health consequences. Adaptation to the new climate reality will not be uniform across different societies. Wealthy, affluent areas in the industrialized West, for example, tend to be greener and leafier—and leisure pursuits enjoyed in natural areas (golfing, for example) can be expensive.

The environmental health aspect of public health has traditionally focused on the harm presented to the environment from pollution and toxins. But it is beginning to flip the question: Could the benefits of connection with the natural world reduce the health gap between the rich and the poor? And if vulnerable communities are denied opportunities to commune with the natural world, are they at even more risk of disease and illness? Poverty is the main driver for mental health issues, as we have seen, which means that people from disadvantaged communities need the ameliorating benefits of nature connection even more than everyone else.

To consider how access to nature could be more equal, we first have to ask: Who owns the land today? Who is it for? In Britain, more than half the land is agricultural (animal grazing and growing crops), and fewer than 200,000 families own two-thirds of the countryside. About 3 percent is common land, and a fifth of land is unregistered on the Land Registry; oddly, and perhaps surprisingly, only 6 percent is built on. Some estimate golf courses take up 2 percent of the land in England, though the figure is hard to verify. Friends of the Earth, an environmental campaigning community, claims that golf courses cover ten times more land than allotments. In the United States, national parks are not spread across the continent: Delaware, Maryland, Georgia, Alabama, Louisiana, Mississippi, Oklahoma and West Virginia do not have any national parks at all.

Who, then, can access the land? People in lower socio-economic groups or from racial and ethnic minorities usually have less access to green space and parks than those who are white and affluent. Evidence suggests that vulnerable populations spend less time in natural areas. In towns and cities, there are fewer parks in deprived areas, compared with affluent areas. Children who live in deprived areas are nine times less likely to have access to nature, through green space and places to play, than children in affluent areas, who may also have access to private gardens. Children living in the poorest homes are six times more likely never to have set foot in a wild open space than those in more affluent circumstances.

The growing disconnection of working-class people from nature has a sorry history. The practice of enclosing land from the British people began in earnest with the passing of over 5,200 Enclosure Acts between 1604 and 1914, which fenced off 6.8 million acres of previously common land and "waste" areas (natural areas that peasants without land could farm). Communities were displaced and often ended up in penury. The nineteenth-century poet John Clare wrote about the painful conversion of the land in many mournful elegies. He deplored the laws for removing grazing areas from those who had no land of their own and for changing the beauty of the landscape—chopping it up into fields where "fence now meets fence." "Enclosure came and trampled on the grave / Of labour's rights and left the poor a slave," he wrote in "The Moors." His landscape in "The Village Minstrel" becomes a place where "Spring more resembles winter now than spring." Through the loss of the natural open spaces, Clare loses his own personal identity and the emotions, comfort and consolation the natural world afforded him as, one imagines, many others without land, money or power did. He was writing about how eco-destruction affected human mental health centuries ago.

On 24 April 1933, a group of around 450 ramblers assembled to make a Mass Trespass on Kinder Scout, in the Peak District of Derbyshire. The ramblers were either factory workers or young, working-class unemployed people during the Great Depression who were fed up with having to ask permission to walk over the open country for their health and well-being.

The mass trespass took place to open up the moors to all people. Led by Benny Rothman, the ramblers defied the law to trespass on the highest point of the Peak District. Six were arrested and sent to trial for riotous assembly and public order offences. In his opening statement, Rothman made the case for the right of access to nature. "We ramblers after a hard week's work in smoky towns and cities go out rambling for relaxation, a breath of fresh air. But we find when we go out that the finest rambling country is closed to us, just because certain individuals want to shoot for ten days of the year. Ramblers are denied the pleasure of rambling over moorland, climbing on the tops . . ."

Sadly, almost a century later, many barriers remain. There is a growing disconnection and it is affecting us all—but it will affect the more vulnerable and marginalized most acutely, from earliest childhood, when access to parks is often dependent on how affluent your neighbourhood is, or whether your parents can afford to pay for forest school in the holidays, to the poorest urban communities who have to cope with urban heat spots without the means to move away. For many who are physically and mentally unwell, it is also difficult to access natural areas, as they become more and more depleted. Our society is only set to get more unequal as the disconnection grows and environmental chaos spreads.

But lack of access to nature isn't the only barrier. In the city of Glasgow in Scotland, there is plenty of green space, and it has increased over the years, but the issue is, as Professor Rich Mitchell of the Social and Public Health Sciences Unit at Glasgow University says, people's willingness to use it. Women, low-income groups and members of ethnic communities tend to report more instances of feeling unsafe in urban parks. Different groups report finding spaces threatening, amid fears of persecution and discrimination.

In an essay for *Outside* magazine, the African-American writer Latria Graham explained how lack of access and racist historical legacies prevented members of her community from experiencing the natural world. In response to a survey by the National Park Service in the United States in 2009, which found that just 7 percent of visitors were African Americans, Graham points to the history of discrimination in public parks and cites Shenandoah National Park's practice of hanging segregating signs that identified PICNIC GROUNDS FOR NEGROES. There were also signs on some bathrooms for WHITE WOMEN ONLY until the 1940s. "People of color are still often left out of the conservation decisions and planning that affect their communities," she wrote in May 2018. "Creating equitable outdoor experiences means dedicating money, energy, and resources to programs that have been denied us for decades . . . It's hard to put a price on learning to read the sky or the ability to smell the wind and distinguish the scents it brings."

In his work with inner-city school kids, David Lindo, who with his Urban Birder project aims to dispel the image problem of the British countryside, observed that Asian or black children can be more reluctant to engage with the natural world at first. Soon enough, they become engaged and interested, but, he said, they may go home to parents who have no interest, and then it's back to square one.

How can society implement and foster nature connection, especially in families which have no experience or background in nature contact? I put the question to Howard Frumkin, the leading physician in the area of nature and health. He talked about tailoring access according to cultural preferences, drawing on his experiences in the United States. "Intentional and mindful programming can either bring people to nature or nature to people. Parkland and tree canopy should be distributed through the city to avoid disproportionate access for some and deprivation for others. And sensitivity to different cultures can help tailor facilities and programmes appropriately. Suppose there's a park in an area of Arab and North African refugees, people who may have suffered a lot of trauma in the process of immigrating. Engaging them in park design may help avoid inadvertent triggers of stress. Many may be observant Muslims who prefer to avoid mixing genders during activities such as swimming. Have a section for women and a section for men. In a Latino neighbourhood, consider replacing the baseball diamond with a soccer field, and provide picnic facilities in recognition of the cultural fondness for family picnics. United States data suggest that white populations favour hiking, so hiking trails would be appropriate for these populations. Parks ought to be for everybody, but we can still aim to accommodate the preferences of local communities."

These aspects must be considered with care and urgently, in order to make a relationship with nature available for all people in our urban world. New evidence from the University of Glasgow suggests the results could be even more powerful than we might expect.

IT'S NOT OFTEN THAT a scientific idea is called "beautiful," but this was how one academic described the keystone work of Professor Mitchell at the University of Glasgow. His research suggests, convincingly, that greener neighbourhoods which offer a connection with nature might actually reduce the health gap between rich and poor and lead to a better, more equal society. "Inequality has barely got any better," he said, at the start of his talk at a conference I attended in Bonn, Germany. "This matters; it's unfair." Could nature connection really reduce socio-economic health inequalities? My ears pricked up.

Mitchell's concept is known as "equigenesis." If an environment is equigenic, it may reduce the gap between the rich and the poor by weakening the link between socio-economic inequality and health inequality. Because of their many health benefits, natural environments are potentially equigenic.

Mitchell spent his early career mapping and measuring socioeconomic inequalities, monitoring the health gap and showing that it was getting worse. It became clear to him that to fix the problem of inequality on a massive scale, "you would need a socio-economic revolution." He began to find the work depressing as he realized it wasn't going to happen anytime soon and, anyway, massive redistribution of wealth and income wasn't democratically supported. "I began to feel that although a lot of that work was useful and valuable, it wasn't moral," he told me. "We were calling for some massive change that didn't seem to be on the horizon and we weren't offering any alternatives."

He began to think about what might be doable on a short timescale. From there, he moved into studying resilience, and specifically the idea that there are groups of people in difficult living conditions who seem to do better than expected. He decided to look at places that were poor, but where the health of the population wasn't too bad. There were certain places in the United Kingdom, some in the north-east, which demonstrated such resilience. There, a sense of community or having something in common was important, such as ethnic identity or common industrial heritage. Another factor was green space.

Mitchell had always been an "outdoorsy" type, and he started following the growing literature around nature and health. He theorized that income-related inequality in health would be less pronounced in populations with greater exposure to nature because green space acts to reduce stress. Looking at populations across England, he anticipated an association between green space and population health, but he wasn't expecting to see such a strong result. "It was a genuine moment of discovery for us," he said. With Dr. Frank Popham, he found that people who lived near parks and woodlands had lower levels of income-related health inequalities. Using national mortality records from the United Kingdom Office for National Statistics, alongside green space measurements, they paired low-income groups with varying levels of urban green space to see the effect on mortality rates. The rates of income-related mortality were much lower in areas with more green space.

Their paper was published in *The Lancet* in 2008 and concluded that it was possible for greener neighbourhoods to mitigate the negative effects of income deprivation on health. The potential benefits of nature connection seemed more powerful to the researchers for people who were from poorer backgrounds and under more stress. One theory was that richer people could reduce their stress elsewhere, through buying access to a golf club or holidays or leisure time in paid-for natural environments. "The local environment may be more important to the people who are less able to purchase stress reduction in other ways," said Mitchell.

Since then, Mitchell and others have continued to chip away at the idea. In 2015, another study led by Mitchell supported the findings. It looked at over twenty thousand people in thirty-four European countries and found that access to nature was the one characteristic that reduced socio-economic inequality in mental well-being (by 40 percent). "If it's true that the potential benefits are bigger for people who are poorer and more stressed, if it's possible to get more of those people into green spaces, the benefits could be massive," he said. Neighbourhoods were like fields in which we grow lives, rather than crops.

But lots of people don't use woodland or open park space, even if it's right next to where they live. Mitchell compared asking some people to start going for a walk in the woods twice a week to asking folk to give up their smartphones. "People would be appalled at that idea," he said. "If that's just not what you or anyone around you ever does, you're asking some people to be deviant in their own communities, which is a difficult thing." To make spaces truly equigenic, said Mitchell, we need to orient people to them, which is what a few individuals are already doing.


I BICYCLED DOWN to a local community garden I'd heard about that ran food-growing and horticultural programmes for children and adults from disadvantaged communities in Basingstoke. It was an idyllic space at the side of a large field, planted and designed so beautifully it had won a number of Royal Horticultural Society awards. On the day I visited, a "Green Therapy" session was taking place, where a group of people with different physical and mental health needs got together every Friday.

The Inspero programme is run by a woman called Catherine Waters-Clark, who was brought up in a farming family in Northern Ireland and had worked in the corporate industry sector. The idea originated when her son wouldn't eat his vegetables. Let's get growing, she said, and saw how much planting interesting vegetables changed his eating habits. "All kids should have a chance to experience this," she thought, and set up Inspero.

In March 2012, a pilot project started with twelve children. "They just loved it," she said. "Being outdoors, planting runner beans, potatoes, they loved coming every Saturday." A site was found, chosen for its location within walking distance of three main disadvantaged communities in the town—Buckskin, South Ham and Popley—to target children from low-income, vulnerable and challenging family backgrounds.

Groups of children worked with the soil, and grew celery, carrots and cucumber and more unusual foods such as cucamelons and tomatillos. The majority of the work with children was funded by Basingstoke and Deane Borough Council, but the funding for the adult therapy had been cut. The session I visited was run by volunteers.

The garden was droning with bees, and my eyes were drawn by the black and white slippers of the broad bean flowers. In one corner, a group of local businesswomen were painting a pagoda-type structure as part of a corporate voluntary scheme. The members of the Green Therapy group were busily digging, planting, being present with other species.

I sat with horticultural therapist Yvonne, who'd first joined Inspero as a client, then became a volunteer and was now an employee. Slowly and carefully, she told me how she'd lost her parents in 2013 and found the bereavement process extremely lonely. She was an only child and also the carer for her parents, who were disabled. "I had to find my own identity because I'd always been a carer, and it was a real journey finding out who I was because I never had the chance."

The garden provided a space for her to think, put things into perspective, come to terms with her loss and work out the problems and issues she had to deal with—both the practical side of selling her parents' bungalow and dealing with personal items, and the act of reimagining who she was. She talked about the fresh air, seeing a red kite circling above, working with the soil, watching things grow and life develop. It was a reprieve from existential loneliness, not only because there were other people in the team who made everyone feel worthwhile and part of a family, but also because of simple things like the sound of a lawnmower on the other side of the road, the people, the community and the plants.

There are numerous brilliant initiatives like Inspero in cities, towns and rural areas across the United Kingdom, and many of the big nature and wildlife charities run successful campaigns to try to encourage people from marginalized communities to feel welcome in natural areas and give children the opportunity to experience the rest of nature. The Sheffield Environmental Movement is one of these, working to provide education and information for Black, Asian, Minority Ethnic & Refugee (BAMER) communities to access the natural environment in the north of England—especially for those who have been severed from their natural environment through migration, urbanization or socio-economic reasons.

If green space is equigenic, if a connection with nature can reduce social inequality and the health gap between the rich and the poor—and Mitchell's evidence for this is robust and highly respected—this is incredibly important. It puts paid to any idea that caring or worrying about the natural environment is a privileged luxury for the affluent. It suggests it should be a priority for public policy decision-making. And it makes the case that, no matter the circumstances of birth, a connection with the natural world, the opportunity to walk barefoot on grass, to plant seeds in the soil, to hear birdsong or touch the bark of an ancient oak, should be a fundamental human right.

But although these initiatives are admirable and many people benefit from nature-based interventions alongside other forms of care, a wider transformation is required. The relationship between people and nature is so deeply damaged, and the consequences are becoming so much more harmful, to the Earth and all its creatures, that a society-wide emergency restoration is what is needed.

# 7

# Ecological Grief

*monachopsis,* noun: the subtle but persistent feeling of being out of place, as maladapted to your surroundings as a seal on a beach.

T HE WORD "ECOLOGY" comes from the Greek *oikos,* meaning "household," "home" or "place to live."

It is all very well thinking about how more connection with the natural world would make people happier and healthier. But there is a problem. How can forest medicine be prescribed when forests are threatened and diminishing across the world? How can people spend time in green spaces when the numbers of parks are declining? How do you have a relationship with somebody who is terminally ill? First, in Western and industrialized society worldwide, we are becoming disconnected from the natural world and overlooking how much we need it. Second, and of course this is related, that natural world is rapidly vanishing; our time on this Earth is haunted by the violent destruction of natural habitats and the extinction of species. Add in the unpredictable and frightening reality of climate breakdown and how it is already leading to the decimation of species—our own included—and our relationship with nature, even if it could be restored, isn't quite as simple as a soothing, serene, Thoreauesque ramble in the wild. The wild barely exists. What is the effect of biodiversity loss on our minds, our inner selves and the collective psyche? Are we, as a collective, harmed by biological annihilation? As the landscape of the Earth changes, how is the landscape of our mind affected? When George, a tree snail and the final member of the *Achatinella apexfulva* species, died on 1 January 2019, did it have an emotional impact that goes further than the scientists that studied him? How do we grieve for the Earth?

ONE OF THE MOST POIGNANT but sadly inevitable words to have gained currency in the twenty-first century is "endling." In the mid-1990s, Robert Webster, a doctor in Georgia, USA, wanted to find a word to describe a patient of his, who was the last surviving member of his family. After a while, he came up with "endling" and sent it to Merriam-Webster for inclusion in the dictionary. He was told that it needed to be used in print a number of times before it could be included. First, he bought an advert in the back pages of a medical journal. Then, he wrote to the "Letters" pages of *Nature* and a discussion ensued with some suggested alternatives ("ender" or "terminarch," because "endling" sounded too "pathetic"), but the word was essentially shrugged off.

In 2001, the National Museum of Australia in Sydney resurrected the word, but used it in a different way. "Endling" appeared in an exhibition to describe the thylacine, a carnivorous marsupial which looks like a cross between a wolf and a tiger, which was hunted to extinction on the island of Tasmania in the 1930s. Artists of all kinds were inspired by the word, and cultural responses that followed included a chamber orchestra composition, science fiction novels, a contemporary ballet, an essay collection and an intense doom-metal album by the band Cull from Portland, Oregon. The way various artists used the word was not, however, as Webster had defined it, but to describe the last animal of an extinct species.

In the 2010s, scientific studies showed that global wildlife populations were falling at an alarming rate. By 2020, wildlife numbers were set to have fallen by more than two thirds. This is called the sixth mass extinction. A study in July 2017, conducted by biologists at Stanford University, suggested that focusing on the passenger pigeon or the thylacine gave the wrong impression that the Earth was *gradually* entering an age of biodiversity loss. Instead, they described the situation as "biological annihilation." The sample of the study was significant. They looked at 27,600 vertebrate species and 177 mammal species and found nearly half of the species had decreased in population size and range. Range is key for species. As living space is reduced, the population will decrease correspondingly, which puts the species in greater danger of extinction. "Our data indicate that beyond global species extinctions, Earth is experiencing a huge episode of population declines and extirpations, which will have negative cascading consequences on ecosystem functioning and services vital to sustaining civilization," they concluded.

One of the consequences of our disconnection from the natural world is that, depending on where we live, many of us rarely witness the depletion of the natural world and biodiversity loss at first hand. But this is changing, as more and more of our natural areas are depleted, often right outside our windows.

The city of Sheffield in South Yorkshire is one of Britain's greenest cities. You can feel it when you're there. The streets are tree-lined, there are lots of parks and gardens and the centre is only five miles away from the glorious Peak District. From 2012 onwards, it has also been the site of horrendous acts of ecological vandalism. A twenty-five-year contract between Sheffield City Council and Amey, a private company contracted to upgrade roads, pavements and street lights, led to the felling of thousands of trees. Over five thousand trees, including oaks, limes and elms, many of them mature, were destroyed. The company planned to fell 17,500 in total. Many locals were furious and organized a protest against the Sheffield tree massacre. In its defence, the private company said it would plant saplings to replace each tree, which showed a galling ignorance about how important mature trees are to our environment, providing oxygen, absorbing carbon dioxide, providing shade in urban heat spots, alleviating flooding, providing habitats for a wide range of animals and making people feel both physically and mentally well.

The protestors, organized by the Sheffield Tree Action Groups (STAG), weren't going to lose 17 percent of their roadside trees without a fight. Skirmishes led to arrests, court cases, detentions, injunctions and one dawn raid known as the "battle for Rustlings Road." But the dendrophiles weren't cowed. They upped their protest game. "Geckos" would stick to walls that were close to safety barriers; "gnomes" would sit and watch from private gardens; "bunnies" would risk arrest by hopping over barriers. The trees and the battle to save them inspired songs, artwork, poetry and a community art group. Residents gave names to many of the oldest and most-loved trees and set up personal Twitter accounts so that people could follow their progress. Wildlife groups mobilized to protest against the killing of particular trees. The Chelsea Road elm, for example, provided a habitat for the very rare and threatened white-letter hairstreak, a butterfly that has a zig-zag of lightning across its wing. It had an added gravity as Britain's elms—some sixty million—have been devastated by Dutch elm disease over the last century. In February 2019, the tree was finally saved, after tree-protectors spent hundreds of hours battling for its life.

Thankfully, the Sheffield tree-protectors managed to delay the cull, and in October 2018 a deal was brokered with Amey pledging to "spend more to retain trees." "In most cases," said Sally Goldsmith of Sheffield Tree Action Groups, Amey are saving the trees with "simple, pragmatic and crucially very cheap solutions—new tree pits and kerbs."

But thousands of trees were already lost, with many residents' everyday experience of daily life altered forever. It had taken many hours of work by many brave people to achieve a result. And this activity, of course, wasn't a one-off. Elsewhere in Europe, one of the last ancient forests in the continent was being destroyed.

OF ALL HABITATS ON EARTH, forests contain the most abundant assemblage of organisms and complex and rich structures. Of all forests in Europe, Białowieża on the Polish–Belarusian border is the largest and best preserved semi-natural ancient forest, the last vestige of the primeval lowland wildwood that covered Europe ten thousand years ago. Wolves, lynx and European bison still live there. In total, the area is 1,500 km, with 40 percent (580 km) spreading into Poland and 60 percent (870 km) spreading eastwards into Belarus. The Polish part is a mixture of managed forest, looked after by the forest administration, and Białowieża National Park, and comprises roughly a sixth of the area, which is protected. Within Białowieża, the "strict reserve" is the largest area on the European continent where nature has been mostly left to its own devices for thousands of years. This most valuable part of the forest is at its core and has been protected since 1921 as a trans-boundary Polish–Belarusian World Heritage Site. You can only enter with a licensed guide, which I was lucky enough to do in February 2018.

Hornbeams twisted and stretched next to tall, slender, light-seeking oaks. Silver birches had fallen next to long-dead limes that dominoed into the perfect moment of decay that would trigger a new stage of life for the lichen and fungi, and send them sprouting and shaping and waxing. Lichen splattered the dull bark: duck-egg blue splurges, as if a child had flicked paint from a brush. Other filigrees were chartreuse-green, bile-yellow, cool-jade and chewing-gum white. Some trees dripped with lichen. On others it looked like fur. The organisms were abundant, I learned, because the air was so clean and unpolluted. In February, it was freezing and I struggled to smell the trees or the earth. Giant clams and spaceships of bracket fungi grew around the trees, fairy stairways to the canopy where globose clouds of mistletoe nested in crooks. Smaller conks of fungi and curled ears of soft brown bracken decorated the fallen ends, splayed around the heartwood. Some fungi were see-through black and jelly-like; others were white, like spilt PVA glue. The pinkish *Fomitopsis rosea* spread like an over-baked cake.

The forest is an important site for scientific and ecological study. It has an abundance of bird, reptile and mammal species: wolf, lynx, fox, badger, otter, bison, elk, roe deer, snowy hare, Eurasian water shrew, as well as 500 species of lichen, 4,000 species of fungi and more than 260 species of moss and liverwort. Modern-day hunters are mainly scientists who live here to truffle out a variety of answers, from the selection process of wolves' rendezvous sites to the habitat preference of the noctule bat; from the details of tawny owl predation to the influence of earthworms on badger densities. Ecologists travel from far and wide to study the long-term dynamics of a regenerating deciduous temperate forest in order to bring the knowledge back to their own depleted landscapes. Others look at the potential medicinal properties of forest species.

Three mushrooms in particular were the focus of promising research by Jordan K. Zjawiony, Professor of Pharmacognosy at the University of Mississippi. He and his team had collected large amounts of *Phellinus igniarius, Phellinus robustus* and *Phellinus punctatus* from the forest. The first two species looked like elephant hooves made from ash or charcoal, and made their homes on the sides of tree trunks. The latter resembled a large brown burn on the arm of a branch, or a spill of caramel. Back at the lab, they dried out the fungi, ground them up and used methanol to extract and examine pharmacologically active secondary metabolites produced by the mushrooms. They found significant immunostimulatory activity, which could act indirectly against cancer as biological response modifiers and inhibit human genes involved in cancer proliferation.

"As the last primeval forest in Europe, Białowieża Forest is the 'treasure chest' that may let us discover new interactions between species, understand biodiversity and have importance for the health of all of us," Zjawiony wrote to me. Of the three thousand plants that are known to contain anti-cancer compounds, 70 percent are of tropical origin, which makes fears that the deforestation of the Amazon—the largest stretch of tropical rainforest on Earth—will increase under Brazil's president Jair Bolsonaro even more acute.

In Białowieża, leaves from various different seasons and years lay softly waiting to disintegrate and turn to mulch and then soil. A fritter of leaf-litter drifted as far as our eyes could see. Underneath the ground, worms nudged their way down desire-lines to feel how the layers were disintegrating through receptors in their wet pink skin. I couldn't hear the chthonic hullabaloo, but I could imagine its crackling, the microorganisms and ants and beetles and soil-dwellers roiling around in the particles of life. The air was cold and the ground was hard and snowflakes billowed around the trees. We kept to the special paths laid out through the forest. The trees on either side were both dead and full of life, literally and figuratively. Rows of hornbeam, lime, alder, oak were more sparsely spaced than I had expected, but even the dead trees had character. One resembled the skeleton of a whale, with a perfect ribcage splayed open. Another bent over backwards like an old lady fainting before the smelling salts reached her.

A multitude of organisms lived and breathed and mated under the dead wood, but the forest was winter-quiet. The deadness of the wood was crucial to the forest's vitality. It had ten times as much dead wood as an ordinary forest, which provided an abundance of ecological niches and microsites. A reddish bank vole popped out of a decayed log like a champagne cork. Enormous piles of dung, each the size of a significant birthday cake, suggested a band of European bison had trudged through the area that morning, heaving bulky shoulders, breathing clouds of hot air into the mute grey dawn. Over the centuries, animals in the surrounding area had retreated into the forest as the people of Europe spread their lives into the lowlands.

But it was so cold I couldn't smell the animals. I couldn't smell anything. It would be so different in a few months. Apart from the occasional generic winter song of a woodpecker and a nuthatch, life seemed to be suspended. Somewhere, there would be wolves: it was their breeding season and the forest was home to two packs. Our guide said that 14 February was the busiest day for mating. Badgers, martens, stoats and weasels were hiding underground from the cold and predators such as wolves and lynx. It was hard to imagine there were more than twelve thousand recorded species of animals in the forest and an estimated twenty thousand, although just knowing it and being there was precious and calming. But it also made me feel sad and angry. After years of protection, the forest was in danger. It was substantially logged in the mid-2010s, and its future is still not secure. It is the ultimate "natural" area in Europe under threat; the woodland equivalent of the Great Barrier Reef. Significant areas have been mown down, leaving just stumps where there were once habitats for warblers and hoopoes and nightjars; the trees, some over a hundred years old, ripped out and taken off, piled high on lorries, to be sold to private companies for firewood, wiping out complex ecological webs of life and habitats.

In March 2016, the then Environment Minister of Poland, Jan Szyszko, altered the forest laws and allowed logging to increase threefold. A bitter conflict ensued. He, and his supporters, argued that a bark beetle infestation was wrecking the forest, and intensive human management and control was needed, including the use of heavy machinery, to destroy areas in order to kill the infested spruces. Scientists, ecologists and activists believed that the bark beetle infestation was just a pretext, and that the minister and others wanted to prove that the forest needed human intervention. "They want people to think the forest can't be left alone, that it has to be managed," a local guide explained. Bark beetle infestations had happened numerous times over millennia, some much more intense than this outbreak. It was absurd and ridiculous to claim the forest was unsafe, the forest scientist and dendroecologist Ewa Zin told me, and simply a ruse to log the forest. Nature will sort itself out, the forest-protectors said. But the Environment Minister, advised by his personal priest, Tomasz Duszkiewicz—who claimed that the Bible said man should "subdue" the land—sanctioned the continuing logging activity, even after the European Commission threatened to fine the Polish government and filed a case with the European Court of Justice.

There is an urgency about protecting a forest like Białowieża. When it is gone, the specific dynamics, the complex interconnectedness of a multitude of different species, are gone for the rest of time. "At some point there will be a collapse, and if and when it happens, it's gone forever—no amount of money in the universe can bring it back," Professor Tomasz Wesołowski, a forest biologist at the University of Wrocław who had researched the forest for forty-three years, said. "With every tree cut, we are closer to this point of no return."

It is a microcosm of what is happening on a much larger scale. Life on Earth is, as far as we know, the only life in the universe. Once it is gone, that's it: life is gone. So ecologists, scientists, activists and NGOs such as Greenpeace mobilized against the logging. With their bodies, they tried to prevent the machinery from being used, which led to violent clashes. Most effectively, they patrolled the area every day and reported their findings back to the European Commission, which eventually found the Polish government guilty of violating European environmental law and threatened to impose a fine. Thousands of trees had already been felled, however; bone, skin and muscle aren't much good against industrial machines. Estimates of the numbers vary between 10,000 trees and 180,000 trees. Plots were logged throughout the forest, apart from in the strict reserve and national park, removing habitats for the rare three-toed woodpecker, white-beaked woodpecker, white-necked flycatcher and species of owl that are protected by EU law. Thousands of trees, hundreds of years old, were wiped out with the signature of a man who didn't understand the science of biodiversity, or didn't care what it meant, as well as the animals that lived in them, fed from them, and breathed the oxygen they produced, and thousands of other species of organism that had lived in various sites since the Pleistocene. It reminded me of the supercapitalists of this world, the oil barons, the climate-change deniers, in positions of immense political and economic power who have wilfully destroyed and damaged the make-up of the planet and the lives of many for their short-term political goals. In Poland, the nationalist politics of a conservative government that despised the trans-national EU trumped the rights of the rest of nature.

When I was there, there was an armistice while both sides awaited a ruling from the European Court of Justice (ECJ). A group of activists were living in an abandoned school and patrolling different zones every day, fearing the logging would start up again, or, worse, planting, after which the land could never return to its natural state. The horror of what had already been wiped out lay heavy in the air. The vocal, pro-ecological National Park Director had recently been sacked. Foresters were preventing ecologists from using the special permits they needed to go to special areas of the forest. But a campaign to interest the population in the forest was working, according to Greenpeace activists. Polls suggest over 80 percent of Polish people support the campaign to protect the entirety of the forest as a National Park. In April 2018, the ECJ ruled that the logging was illegal, but in 2019, lawyers and environmental organizations such as Greenpeace and the World Wildlife Fund (WWF) sounded the alarm about new plans to restart commercial logging before the wounds of the forest had healed.

Journalist Adam Wajrak first visited the forest on a school trip, fell in love with it and moved to the village in the late 1990s. I asked him about his favourite parts of the forest over cake in a hotel restaurant. He described a dynamic area in which he'd spent time with woodpeckers. One day, during the intensive logging, he returned to the exact spot. "I couldn't recognize the place; it was smashed," he said. "This little path had become a muddy highway with huge holes from the tyres, all the trees were cut. I couldn't recognize the place, everything was so upside down." He returned day after day with his GPS to find the exact place where the woodpecker was, but he couldn't work it out. The forest was completely unrecognizable. He compared it to returning home after a bomb has destroyed your house and village, with no sign of where anything had been. He was disoriented and confused. "I wanted to cry, I was fucking angry," he said.

THE ECOLOGICAL FUTURE IS also dependent on the changing climate and how it will affect landscapes and populations. Our restorative environments—and thus a significant factor in our mental health—are dependent, partly, on weather, which is currently in flux.

In Sweden, climate-change projections are for warm winters with less snow and colder, rainier summers. A study of antidepressant use between 1991 and 1998 by the psychologist Terry Hartig suggested that mental health problems might be affected by decreased opportunities for nature restoration. He found that rates of SSRI prescriptions increased for both men and women in an unseasonably cooler month of July. If people can't do as many of the outdoor activities as they're used to, in order to, using Hartig's phrase, ameliorate psychological wear and tear, we can imagine that the incidence of stress and mental health problems will rise. Emerging evidence links hotter temperatures to increased mental health problems, illness and suicide.

In other areas of the world, the changes of the Anthropocene are also being felt.

Venice, a tiny town at the bottom of the Plaquemines in Louisiana, and the last settlement on the Mississippi River, is already feeling the profoundly disruptive effects of climate change. Flooding and coastal erosion have led to the loss of homes and communities, and the marshlands which once provided a buffer between the town and hurricanes slipped into the sea. Over the 2010s, the population halved as people moved north. In April 2018, Plaquemines was one of six southern areas in Louisiana given a $40 million federal grant to boost flood resilience. It chose, as one of its programmes, a treatment project for mental health problems and substance abuse. The psychological toll on people living in climate-vulnerable places is just starting to be realized and, one imagines, it will require significant attention and funding worldwide as the climate outlook and its effect on public health worsens.

In Kulusuk, Greenland, where the ice has melted, rates of depression, suicide and alcoholism have risen. In the Nunatsiavut region of Labrador, interviews with residents by public health researchers found that the melted ice and changing weather patterns made people feel stressed, depressed and anxious. It strikes at the very heart of identity. "Inuit are people of the sea ice. If there is no more sea ice, how can we be people of the sea ice?" said one resident to the public research team.

The Inuit of Baffin Island in the Canadian Arctic use the word *uggia-naqtuq* to mean "to behave strangely, unpredictably, or in an unfamiliar way." Their homes are changing, what they see every day is changing, the weather is changing, what they can hunt and eat is changing—and it is starting to change the people themselves. It is certain that, as climate chaos accelerates, many—if not most or even all—countries will be affected in some way. What happens in the Arctic doesn't stay in the Arctic, it affects the rest of the planet, as the National Oceanic and Atmospheric Administration's Timothy Gallaudet has said.

Climate change will affect our mental health, then, through various pathways. It will directly expose populations at the front line to trauma, such as floods, vector-borne diseases and extreme heat, loss of homes, loss of life, loss of health and loss of ways of life and cultures. Farmers in the Australian Wheatbelt, whose farms have blown away in dust storms, have compared losing their farms to a death. "It's terrible to know that the soil has been there forever, since the beginning of the Earth, and your greed and mismanagement makes it blow. It's a really horrible thing to see, and I hate seeing it on other people's farms," said one.

It will disrupt our communities, identities and social environments in more subtle ways than we can yet imagine. This effect has been called "slow violence." It is going to lead to mental health issues and a sense of distress and disorder. It has already. As Glenn Albrecht, the Australian academic who coined the word "solastalgia," points out, what's disordered isn't the ecological grief and anxiety, but "the world that is causing you to feel that way." It is a natural response to loss—and it is likely to become more common.

HOW HAVE WE GOTTEN into this mess? Well, that is another book. But in brief, it is our economic systems, our obsession with infinite growth and the way we perceive the world around us, with man at the top of the hierarchy and the world at his disposal. The short-termist nature of human thought limits us from seeing how our actions could lead to catastrophe for future generations. It is hard—not to mention irritating—to imagine that eating a burger, or using shower gel with microbeads in it, or buying fruit in plastic containers, or going on holiday by plane, contributes to planetary damage. Most of us don't know that, for example, tea bags have plastic in them. Our eyes have been distracted from the great prize of Planet Earth by rapacious consumerist culture, insidious advertising and media that persuade us we can't be happy without this or that.

Collective amnesia is also at work. It is easy to be distracted from nature by our busy lives, by social media, work and technology. It is easy to forget that we are part of nature, and we only breathe, eat and drink because of it. We choose to bask in screens instead of mirror-calm lakes, burbling streams and, above, starlings, swallows and buzzards. "We are involved in a kind of lostness in which most people are participating more or less unconsciously in the destruction of the natural world, which is to say, the sources of their own lives," wrote the farmer and visionary writer Wendell Berry. "They are doing this unconsciously because they see or do very little of the actual destruction themselves, and they don't know, because they have no way to learn, how they are involved." It is a kind of unconscious planetary suicide.

But how do we turn it around? Grief can be paralysing, but it can also spur us into action. Perhaps grief is what we need to change course. I worry about the long-term future of my daughter and her generation most days. I can relate to the BirthStrikers, a movement of women and men who have decided not to have families because they are worried about bringing children into a world of environmental collapse and wide-spread political inaction. This is now, as Alexandria Ocasio-Cortez, a US Congresswoman, has said, a "legitimate" consideration for young people. When I think about my daughter's future, I recognize that we are extremely privileged to live in a wealthy country which is less likely to be affected by climate instability than other countries. I am also middle-class and could consider, for example, moving to an area with less air pollution or extreme weather events. Many people are not as lucky.

How do parents prepare for or even conceive of an unstable future for their children? I am trying to figure it out. But in the short term, I will give my daughter the opportunities to commune with the natural world. I will counteract the hours she'll spend in the classroom by taking her outside. I will try to instil in her a sense of awe and wonder in birds and wildflowers, as my parents did with me. And I will do what I can to make a difference.

PART V

# BARK

∽

# 8

# The First Primrose of the Year

When I trace at my pleasure the windings to and fro of the heavenly bodies, I no longer touch earth with my feet. I stand in the presence of Zeus himself and take my fill of ambrosia. —PTOLEMY

I am shedding my skins. I am a paper hive, a wolf spider,
the creeping ivy, the ache of a birch, a heifer, a doe.
—VIEVEE FRANCIS, "Another Antipastoral"

I am conscious that shivers of energy cross my living plasma from the tree, and I become a degree more like unto the tree, more bristling and turpentiney. —D. H. LAWRENCE, "Pan in America"

Ancient trees fill us with awe, and perhaps, in an increasingly godless age, they occupy some of the vacant space in minds once filled with religion. —PETER MARREN, "The Ramsbury Elm"

AND SO I SAT, that topsy-turvy summer, with my baby daughter playing in the soil, fearing for what we're losing on a spiritual level, as well as for our bodies and minds, as we overlook or ignore the benefits of nature. The health of the planet seemed to be slipping through our fingers, with rapid species decline, habitat destruction and climate chaos. I saw how our disconnection and the consequences thereof were leading to global, planetary despair. Even if we're unconscious of it, is some part of our spirits afflicted by the mass burglary humanity has committed on the Earth and the rest of the living world? Was the dissonance of taking and receiving life from Earth while treating it like shit causing our spirits harm? I know that I feel rotten and out of sorts when I am selfish or hurtful to the people around me.

While I was adapting to the vulnerability of early motherhood, with the added worry about the kind of world I had brought a child into, the garden, the green shoots, the tiny golden millipedes, the banded snails, the goldcrests, the willow tree all gave me a foothold on which to do it. I planted hundreds of seeds and manically filled the house with seed trays and pots and old bowls filled with seedlings. Putting a seed in a thumb of compost and seeing it poke its green flesh through the soil and unfurl was soothing. It was one place that felt constant and safe and not subject to interpretation or chance. When everything around me felt subject to change—too new, too frightening, too uncomfortable—and I felt like I was living in another body, in someone else's life, the garden grounded me. The soil was turning, things were growing, worms were pushing themselves through the soil and the garrulous blackbirds were making their nests and their feelings known.

But the weather patterns that year continued to feel off and, at the same time, more news of climate breakdown and mass extinction were feeding into a sense of dislocation and worry. The winter of 2017–18 in England had felt brutal—colder than average, with lots of rain. In early May we still had the heating on and it was just 2°C at one point. Apart from a few unseasonably hot days in April, it had felt hard, cold, dark, glum and interminable. The sun barely came out for one ten-day stretch. The blossom on the apple trees was six weeks later than it had been the year before, and there were no swallows or swifts to speak of in the first week of May. Without the expected transition to spring, I felt suspended. Was the world still turning? It often didn't feel so. The world lurched from freezing to boiling for a few days and then back to freezing. I was seasonally confused, whiplashed by the weather. Where were the swifts?

You could always count on the seasons living in England, almost like clockwork. The warblers would come back at this time, the swallows swerve in around that day. Knowing that every winter will change to spring, especially in a usually seasonal country, has cheered people up since Tennyson's time and presumably before: he wrote, simply, "I can but trust that good shall fall / At last—far off—at last, to all, / And every winter change to spring." Indeed, seasonal change was at the very heart of Tennyson's striving to find hope in grief at the loss of his close friend Arthur Henry Hallam.

I was so relieved when the sun came out and it felt warm again. I sat in the garden with my daughter and we watched the chickens. I was beginning to emerge from the shock of new motherhood. The challenge of a sometimes frightening, eclipsing love, the strange pressures of modern parenting and a feeling of being pushed and pulled between her wants and mine—sleep!—had rocked my emotional core. But the garden and the returning spring told me, profoundly, that the hallmark of life is change; that I was just adapting to a different phase. Above where we sat, a wren jinked around the foliage. After it ducked into the ivy, I heard a drone and watched a bumblebee visiting various bells of lungwort, a name that belies its pretty petals of pink, purple and blue. The bumblebee had two panniers of bright yellow pollen hanging around its legs. To our right was a bee-fly, sticking its proboscis, which is around the length of the pear-shaped body, into jasmine flowers to find nectar.

Much in the news around that time felt random and cruel: a terror attack at a pop concert for young girls and boys; families drowning in the Mediterranean; the crisis in Syria; the Rohingya genocide. But the natural world didn't. Of course, natural disasters, disease, freak accidents are also part of nature, but there are some elements of constancy that one cannot find in the human world. Buildings do not grow again, neither do the departed, or kings and queens and spiritual leaders. Tuning in to the rest of nature's cycles can offer a tempo and rhythm in an often rudderless modern world. The return of migratory birds. The constellations of the sky. The shape of an oak leaf. The stripes of a badger. There is a soothing constancy to nature. And of course the transitory nature of the seasons— the changing trees, the behaviour of birds and animals, the turning of the globe, the cycle of life in one year—can also be a reminder that time passes and things heal. For now.

I don't follow any particular religion or spiritual path, but I feel at my most spiritually well (by which I mean, my inner voice is calm, peaceful, balanced) when spending time in the natural world. For me, this is found by swimming in rivers near my house, among the rotifers and kingfishers, the damselflies and water mint. I go to "church" in the rivers and lakes and the sea. I can only describe it this way: the essence of me is at peace in some deeply primal way. It is a primordial feeling, which I think many of us experience when we go camping or swimming in rivers and lakes or walking through forests and woodlands. Perhaps it's akin to what thrill-seeking selfie-takers are searching for when they photograph themselves next to a waterfall. Or why we watch animal horror films and survival thrillers. Or sign up for physically gruelling endurance experiences such as Ironman or Go Ape. The popularity of such tests and activities suggests that many of us crave and need that feeling, and perhaps it has replaced the role of religion in our increasingly secular society. But how will we access that feeling of a primordial state today, when the connection with the natural world is in a downward spiral? Where will wild swimmers get their fix now that pollutants are so widespread that no river in England is certified as safe to swim in?

At the beginning of my journey, I'd regarded ecopsychologists who spoke of a planetary-related mental health crisis with scepticism. I'd thought the idea of a collective grief and despair at the planet being in crisis was kooky and off-the-wall, but I was starting to think there could be some truth to it and it warranted serious thought.


ECOTHERAPISTS AND ECOPHILOSOPHERS propose that we have moved so far from our love of the natural world, our biophilia, that our emotional lives have been harmed. It's an idea shared by others in the philosophical movement led by Norwegian philosopher and mountaineer Arne Næss called "deep ecology," which calls for a deep understanding of the interconnectedness of life on Earth and the value of biodiversity, in order to change societal systems on a fundamental level. Deep ecologists call for a completely altered attitude to other species.

Ecopsychology pioneer, psychotherapist and social activist Chellis Glendinning diagnosed Western culture as suffering from "original trauma," caused by our severance from nature and natural cycles. The symptoms, she wrote, are recognized symptoms of post-traumatic stress disorder: "hyperreactions; inappropriate outbursts of anger; psychic numbing; constriction of the emotions; and loss of a sense of control over our destiny"—all of which, she argued, we have come to accept as normal. And yet we have no language with which to address it.

Though Glenn Albrecht, the wordsmith we first met in the Introduction, is forming a new vocabulary of "new words for a new world" to describe our emotional responses to a changing planet, there is a paucity of language in this area which does us a disservice. "Green space," for example, is how the natural world is described in most scientific studies, but isn't it just the most boring and dull phrase? "Nature" itself is problematic as a word because, of course, we are part of nature even if we don't think we are, or accept we are, so in a way it solidifies the separation between people and the rest of the living world. "Natural capital," the term used by the British government to refer to the wealth and value of "ecosystem services" (another stinker) is horrible and highlights an attitude that needs to change: that nature is only worth something if it is giving us dividends. "Environment" doesn't even sound like something that's alive. "The environment," rather than "our environment," underlines the misconception that we live on some kind of separate planet, that we are somehow not alive because of the living world around us. "National park" sounds like something for us, rather than homes for species in their own right. Even the most beautiful natural areas of England are given the most atrocious names. The marshes near my London home were an SSSI, a Site of Special Scientific Interest. The ugly acronym doesn't do justice to the rich ecological habitats that thrived and hummed there. It doesn't convey the wonder of hearing the trill of a cricket rubbing its wings together or a kestrel hovering above wildflowers.

These words and terms frame nature as a lesson, or something that is ours, and strip away the wow factor and the wonder. We need a new vocabulary, to describe both the peril and harm we have enacted on the Earth, but also the bliss and the soul-soothing that a relationship with nature can offer. "Ecocide," for example, with its acknowledgement that humans are complicit in the destruction of the Earth, seems to me a much more realistic and authentic word than "environmental collapse." The "Symbiocene" is a worthy alternative to the "Anthropocene," the word used to describe the epoch of time we live in, defined by human dominance of the planet, as it underlines the interconnectedness of all species and biological processes and life, something we seem to forget.

"Climate change" is too innocuous a description of what is happening to the planet. "Extinction" doesn't say anything about humanity's complicity in global trends. Even referring to pigs as "pork" or cows as "beef" emphasizes our alienation and disconnection from the land and other living creatures. (My own personal kickback is to spell out and thank exactly what food is on my daughter's plate. If she has fish fingers and chips, we thank the fish and the potatoes, and the plant and the soil and insects and worms that make it possible for them to grow. If she has sausages, we thank the pig for its life. If it's honey on toast, we thank the bumblebees and the flowers and the wheat and the soil and so on. My father gently reminded me, when I told him, that bumblebees don't make honey, which is Robert Pyle's "extinction of experience," referred to in the Introduction, in action.)

Even the medicalization of nature—the words "ecopsychology" and "ecotherapy," the idea that we need to "make the woods work" for us—demonstrates that we still see ourselves as takers and overseers, the authority figures, rather than being on an equal footing with the rest of nature. A shift away from the extractive way of treating the earth—subduing and dominating it, pillaging and using up the land—would help.

How we talk about the problem, and the solution, weighed heavily on my mind. I also wondered if psychotherapy or ecopsychology could help us with the growing climate dread and planetary despair, or if it was too late. Was there a danger that, as people wise up to ecocide and how human activity is wiping out other species, the guilt will be so paralysing and extreme that it could lead to mental health problems or widespread spiritual sickness? How would a therapist even counsel someone with planetary despair when the upshot does not look good?

Mary-Jayne Rust, a Jungian analyst and psychotherapist who was one of the earliest ecopsychologists practising in Britain in the 1990s, has spent decades thinking deeply about the subject. Her work suggested the ecological crisis was never far from her mind. "The very thing that is causing our crisis—over-consumption—has become our palliative, to soothe away our anxieties about the damage we are doing to the world," she wrote. "Some people liken this to the vicious cycle of addiction."

As well as conducting therapy outdoors, she is a specialist in eating disorders and an art therapist, and had started her career at the Women's Therapy Centre, where she was part of the early feminist psychotherapy movement. I travelled to meet her at home in leafy Highgate to learn how we might live psychologically healthy lives in a depleted world—and how we have ended up at this point.

If someone brings their concerns about what is happening to the

Earth to you in a therapy session, when your objective is presumably to make someone's life easier, where do you go with that? I asked. She pushed back on my assumption. "My objective is to listen, very actively, to help people explore unhelpful patterns of thinking and behaviour that they get stuck in. In a way it's to deepen, which in my experience can help someone to become more satisfied, and develop a sense of meaning. This hopefully brings a deeper happiness in the end, rather than the shallow happiness we are sold by Hollywood."

Rust suggested that we need a "Me Too" movement for the Earth. "That's what we see with the individual in psychotherapy, you beaver away for some time and then"—she clicked her fingers—"suddenly you see another change; when all the smaller changes add up, then there is the possibility for transformation."

This made me think of an argument of the academic Nicole Seymour, who has written about the "heteronormativity encoded in dominant depictions of nature." I wonder how much our presiding interpretation of nature as an element *out there* to be subdued and/or romanticized has led to a one-dimensional and less complex emotional response. All the natural history writers and poets I grew up reading were straight, white, educated men. I'm hungry now for other stories and other perceptions, different narratives. Perhaps it's time to move on from the traditional way we have been encouraged to think about the natural world, via Wordsworth and Thoreau, and start again, working out how nature makes us feel, revelling in its—as Seymour puts it—"messy grossness," and finding new words for the relationship. Perhaps, as other barriers to the way we think about life break down, it is time for a new cosmology, a new love story.

Certainly, the human–Earth relationship is going to have to evolve. But what would it take to remove our heads from the sand? What would it take for the media to read a scientific review that said that all the insects will be gone in a hundred years, which, of course, means all the animals, including humans, will also die, and put it on the front page? Rust thought we were in a liminal space. "There's all sorts of things in the pipeline . . . Hopefully there will come a time when we say to ourselves, actually we've *really* had enough now! Just like the change in an individual, this cultural shift can appear to happen quite suddenly, such as the Berlin Wall coming down. Yes, people are beginning at last to protest on the streets now." When we met in 2018, I wasn't sure we were there yet, but Rust was prophetic. The following year, in May 2019, the UK parliament, following demonstrations in London from Extinction Rebellion, the climate breakdown protest group, declared a "climate emergency."

ONE MOVEMENT IN WHICH nature is incorporated into daily life to satisfy a spiritual need is Druidry. In Britain, there has been a revival and resurgence of interest in this spiritual path which supports the idea that people are suffering psychologically without a connection to nature. I wondered if it might offer a framework for thinking about how the relationship can affect mental, emotional and spiritual health. The Chief Druid of the Order of Bards, Ovates and Druids is a trained psychotherapist and psychologist, so I travelled to Lewes in East Sussex to meet him.

Lewes is a picturesque town most famous for its cheeky, carnivalesque Bonfire Night celebrations. It has a castle, its own currency and plenty of artisan coffee shops. I met the Chief Druid, Philip Carr-Gomm, at a trendy independent cinema in an old depot, and we headed out onto the South Downs. The hedgerows were white with suds of hawthorn as we climbed the hill, the sea to our right, Mount Caburn looming inland, and the Downs stretching on and on. We passed a golf course and quickly found ourselves in a huge expanse with 360-degree views. Red kites hunted around sheep and lambs. Skylarks hovered high in the sky like frozen Frisbees, bleeping like fax machines.

Carr-Gomm has a mane of curls, wide, friendly eyes and a big smile. He studied psychology at University College London and then trained as a psychotherapist in psychosynthesis (an offshoot of psychoanalysis which includes a spiritual dimension). Now he is in his mid-sixties, a father of four, and has an active presence online with his Monday "Tea with a Druid" on Facebook regularly attracting thousands of people. Since leaving private practice as a therapist, he has continued to write books, give talks and run outdoor retreats.

We discussed how our relationship with the natural world was missing from the whole field of psychology, with risky consequences. "You can sort out your relationship with the deity, the transpersonal, your fellow human, your sense of self and your consciousness, but until you actually also have a deep and meaningful relationship with the natural world, you'll never be whole or complete," he said.

"Are mainstream psychologists doing people a disservice by ignoring that element?" I asked.

"As soon as Freud opened the box about sex, for us looking back it's so obvious it's such a powerful driver that to not engage with it is just nuts . . . It's the same thing with nature. There will come a time when people will say it's so obvious. How come we were so self-obsessed that we didn't think about all this?" he said, pointing to the Downs around us.

How can nature soothe modern mental health problems, and how had he approached that as a psychologist?

"At one level, the reality is that we're not alone and we're not even single. At all these different levels, there's this extraordinary community of cells, and being in nature reminds us of the fact that we're not alone. We're here with all this stuff," he said, again pointing to the shrubby copses and lambs.

What did it actually look like, in his experience? There are two facets of Druidry that combine a relationship with nature and good mental health. First, temporality: Druids tend to celebrate eight solstices and equinoxes through the year. They pause and pay their respects. It sounds so different to fast-paced modern life. The other facet is space. Druids have a strong attachment to sacred space and special places. They travel to these spots at different times of the year.

Carr-Gomm pointed out Mount Caburn in the distance, one of the highest landmarks of the South Downs and the remains of an Iron Age hill fort. As part of one of many excavations, pollen records showed that yew woodlands were there, or planted there, around 2000 BC, perhaps by Druids. "In urban life you don't even know what time of year it is because you go from one box to another, wake up, open a box of cereal, go into your metal box, spend all day in your glass and steel box, looking at a box; when you get home and you're knackered and you've whacked a little box in a microwave, you turn on the box and eventually get carried out in a box." Marking time and space in the natural world was an appealing counterpoint.

·   ·   ·

THE IDEA THAT WE are cauterized from nature, uprooted from the land, having ripped out Mother Nature's umbilical cord, and are suffering the psychic consequences is not a typical topic of contemporary psychological discourse. When I mentioned my fear of bringing children into a world of climate change and ecological breakdown to my CBT practitioner in 2018, she looked a little startled. When I did the foundation year of training to be a psychotherapist in London in 2014, there was no mention of the rest of nature. Our environment, and our separation from the wild, did not figure in the slightest, anywhere. Which isn't to say it is completely absent from the psychoanalytic canon.

Carl Gustav Jung, the Swiss psychiatrist and psychoanalyst, is remembered most of all for his concept of archetypes: the universal, primitive, innate patterns and tendencies in the collective unconscious which mould behaviour, thought and consciousness. He was also a farmer and lover of the natural world. His landscape was the thick spruce green of the Black Forest, the high Alpine peaks of Italy, the deep chasms of the Rhine and the crests of the Jura Mountains. The woods, he wrote, were the place he felt closest to the meaning of life, and its "awe-inspiring workings."

Less well known is his writing about the human relationship with the rest of nature and our own origins. He advocated spending time in nature as crucial to a healthy psyche. Writing to a colleague in 1947, he said, "You must go in quest of yourself, and you will find yourself again in the simple and forgotten things. Why not go into the forest for a time, literally? Sometimes a tree tells you more than can be read in books." He also said that people in middle age were more in need of an "experience of the numinous" than the young, to help them navigate the second stage of life. The word "numinous" is derived from the Latin word *numen,* which means the divine will, divine sway or power of the gods, or a divinity, godhead or deity. Jung was referring to an experience with a spiritual or supernatural quality.

Jung believed that we had forgotten that we are primates, and that we need to make allowances for primitive layers in our psyche. He employed the image of a house to explain.

Imagine a house with two storeys. The top floor was built in the nineteenth century, and part of it (the windows, etc.) renovated in the twentieth. In the early twenty-first century, a loft extension was built on the roof. The ground floor was erected in the sixteenth century, but the original stonework actually dates back to the 1100s. The cellar is Roman, a cool and cavernous space with a dusty floor. Underneath the cellar, there lies a small cave with a variety of Neolithic tools and trinkets. In the lower layer, there are fossils and bones from flora and fauna, some of which are extinct.

The house represents the human psyche. We live on the top floor, perhaps in the loft extension. We know little of the ground floor, but we know really next to nothing of what's under the ground. "Of that we remain totally unconscious," wrote Jung. However, even though the lower storeys are buried, they are still alive in our consciousness. The upper storey, where we live, is influenced by its foundations, while the lowest layers of our psyche still have an animal character. What we searched for during our evolutionary past—in the natural world—formed and moulded the very structure of the human mind. As the biophilia hypothesis goes, humans spent millennia of their evolutionary history driven by the rhythms of the natural world—where to find the nuts and seeds, when the animals would appear at night, how to find a safe shelter to sleep—in a constantly changing, dynamic environment.

I wondered whether, for example, when my friend said being in nature made her feel "whole," and when I had experiences in which I felt completely connected to the landscape, almost as if I were in the ground, the trees and the rivers, it was this sense of a primordial state, of something deep, pre-conscious and pre-verbal in the areas of the brain or the mind that still have a primitive, animal element.

I occasionally have moments in the natural world that are only really comparable to hallucinogenic drug-induced experiences—when time stands still and I enter a different kind of consciousness, an ineffable state of transcendence. I have it mostly when I'm swimming in rivers—ideally naked, mud between my toes, bugs in my hair, the sound of birds and a waterfall, and the sun dappling through the trees. Suddenly, I am amorphous and enter a different dimension. I feel as if the water is a portal into another world, that the sap from the trees is filling my veins, that I am dissolving into the water—and yet I am more myself than I ever have been.

Have we lost this sense of the metaphysical in nature, the numinous? Is it a risk, in fact, to turn our backs on the wild places which feed our spiritual selves? By setting up our lives in a prosthetic, man-made world, have we actually endangered ourselves? Humanity has never faced a psychological environment like this before. Could it fundamentally change the human spirit?

The severance of people from the natural world was, in Jung's view, a disaster, and led to a loss of balance "on all levels," cosmic and social isolation and psychic injury. He called it a loss of the "bush-soul" and laid the responsibility at the feet of "more than a thousand years of Christian training," which he saw as an attack on the natural joy found in nature. He believed that modern humans living in the West had fallen into psychological shock and profound uncertainty following the First World War.

Jung also believed that we were being swept into a future which would be even more violent, the further we travelled from our roots. Little did he know at the time that humans in the 2010s would start to experience the violence of climate change, from the worsening of severe storms such as Cyclone Idai to the increase in wildfires, extreme heat and drought across the world.

Jung's solution was that every person should have their own plot of land to allow the primitive instinct to come to life again. Most people, of course, don't own their own land, or even have access to an allotment these days, but one can imagine Jung recommending community gardening if he were alive today and scolding local councils with long waiting lists for allotments. In 1956, a few years before his death, he wrote a letter extolling the virtues of growing plants, and described it as "the marriage of the human psyche with the Great Mother."

I wondered whether this loss of balance Jung referred to had also affected our attitude to the later stages of life. Faith, religion, spirituality have all helped people cope with death. If a connection with nature can offer transcendence, can it also help us at the end of our lives?

# 9

# And in the End . . .

The profusion of stars told him unambiguously that he was doomed to die, and the thunder of the sea only yards away—and the nightmare of the blackest blackness beneath the frenzy of the water—made him want to run from the menace of oblivion to their cozy, lighted, under-furnished house.                    —PHILIP ROTH, *Everyman*

The grand doctor had said that he must have fresh air and Colin had said that he would not mind fresh air in a secret garden. Perhaps if he had a great deal of fresh air and knew Dickon and the robin and saw things growing he might not think so much about dying.

—FRANCES HODGSON BURNETT, *The Secret Garden*

IN 2017, shortly after I became a mother, I thought about death a lot. Our daughter had a benign but frightening condition called blue breath-holding spells, whereby she would occasionally struggle to catch a breath while crying, turn blue, then deathly pale, and lose con-sciousness. It looked as if she was dying and, even after we'd seen a doc-tor, and we knew what was happening, it was rattling. It took over a year before I wasn't worried every day that she might die in my care. I strug-gled to adapt to the reality that I had given birth to a life, my great love, but also a death at some point. I wondered whether my new-found fear of death, and the fact that it felt utterly taboo to speak a word of it to any but my closest confidantes, had something to do with my environment, the society I lived in, and its disconnection from the natural world.

Psychologists have suggested that human consciousness of death can lead to an ambivalence towards nature, a recoiling away from our creatureliness. Rejecting our animality, as the American environmentalist and writer Paul Shepard put it, may evolve into a "rejection of nature as a whole." Partly this is because Judaeo-Christian teaching, which has influenced so much of our thinking in the West, decrees that man is made in the image of God. We can hardly, therefore, be seen to be animals! From an early age we are taught that we are not animals, in the same way that other creatures are. The interconnectedness of all life is brushed under the carpet in an effort to keep human beings at the top of the hierarchy (and justify the extractive philosophy which allows us to remove resources from the land and destroy habitats with no thought of the consequences). But the urge to withdraw from nature and its untameable natural forces—decay and the fragility of life—might well be part of the drive to deny death.

In the early twenty-first century, psychologists at the University of Groningen in the Netherlands asked a group of Dutch students to report how often they thought about death in different environments—when they were in the wilderness, a managed natural environment and the city. They were also asked how often they thought about freedom. Sander L. Koole and Agnes van den Berg found that 76.7 percent of the students were more inclined to think about death in the wilderness than in managed nature, and 68.9 percent thought more about death in wild nature than in the city. Most of the students lived in urban environments, where they would have been used to threats and dangers to their lives through crossing a road, for example, rather than the more abstract threats in a wilderness area.

But they also thought about freedom when in the wild, natural areas, suggesting an ambivalence of emotional experience. About 81 percent were more likely to think about freedom in the wilderness compared with a cultivated natural area, and 77.8 percent were more likely to think about freedom in the wilderness compared with a city.

The researchers hypothesized that terror management theory (TMT) could explain what was happening. The theory's premise is that people have a fundamental need to protect themselves from the idea that their own death is inevitable and ultimately out of their control, and the anxiety that this thought creates. The team saw the ambivalence and avoidance of nature as a "terror management process." Wild nature seemed to remind most people about their death, and the existential terror related to it, and led to a shunning of the natural environment. They also found that the visual preference of people who preferred wild settings over managed settings could be "weakened" by reminding them of their mortality. Perhaps, as Richard Mabey has suggested, it is our awareness of death and the consciousness of mortality that makes us "yearn so much to leave our mark on Earth."

It goes some way towards explaining our separation from nature, how and why we have turned away from the natural world and what we notice, or pretend not to notice. I believe that this partly explains what is happening in our wider society in response to ecological collapse. It mirrors the society-wide act of ignoring the scientific fact that our world is dying. We are very able, seemingly, to read about mass extinction, or accept that we are consigning *Homo sapiens* to be a short-lived species, but at the same time continue our lives as if nothing has happened. We really don't want to believe it and we have an incredible capacity to turn away, move on and distract ourselves. By distancing ourselves even more from the living world, we can continue to put our fingers in our ears and hope it will all go away or that someone else will fix it. It is the denial of death on a global, planetary scale.

An experimental psychology study in 2001 by Jamie L. Goldenberg and others into animals, disgust and mortality supported the findings of the Netherlands group. The study hypothesized that being animals, and being conscious of our creaturely nature, was threatening and unsettling because it reminded us that we are mortal. It found that people who were reminded of death showed an increased disgust towards animals. The more aware people were of mortality and death, the more they wanted to read about humans being distinct from animals.

It appears that, through our disconnection from the natural world, we are also more separated from death and disease, and therefore less able to cope with them. Is our treatment of the elderly in the United Kingdom—a place where over a million people over the age of seventy-five can go more than a month at a time without speaking with a friend, family member or neighbour—related to our fear and disgust at the ageing process, a symptom of our inability to address death, and a reason why many of us blanch at aspects of the natural world, be they spiders or foxes or female body hair?

What is the alternative? Suppose, as the American anthropologist and writer Loren Eiseley wrote, we see ourselves burning like maples in a golden autumn: "That we could disintegrate like autumn leaves fret away, dropping their substance like chlorophyll, would not our attitude toward death be different?" As John Scull found in his ecotherapy sessions, could the metaphors of nature help us reconceptualize the ageing process with greater kindness and respect? Can a relationship with nature soothe someone in the final chapter of their life?

DUNGENESS IS THE TIP of a shark's tooth, jutting out from the lip of the south coast of England into the English Channel. In December, the sea foams like the dregs of a pint of warm bitter. Seed heads and sea kale and sea cabbage are blown bone-white by the wind and the sun. The coast is undulating waves of shingles and pebbles. The gulls and lichen are the only visible living things in winter—tenacious life in this flat, blown desert plain. The sky is the thing: pale pink, ultraviolet, charcoal grey, silvered. We are surrounded by it. The horizon extends and calms and refreshes—there is so much of it.

Noses red, we returned to the cottage. From its back window, we could see the nuclear power station. At times, it emits an apocalyptic roar which lends Dungeness the dystopian feeling it is famous for. A couple of houses down, and one of the reasons why people make a pilgrimage to this elemental desert—the only desert in Europe—is the cottage that was owned by the late Derek Jarman, the film-maker, writer and counter-cultural figure. Prospect Cottage is black and yellow, like a fat bee. In the front is Jarman's famous garden, which he cultivated, wrote about and continued to tend while he was suffering from AIDS-related illness, which killed him in 1994.

*Modern Nature* is the title of Jarman's published journals, written over the last couple of years of his life. They are part gardening journal, part early childhood and adolescent memoir and part celebration of love, friendship, sex, colour and art. I wanted to examine Jarman's writings to see if his garden, his relationship with the rest of nature, had made dying easier. Death was already tragically present in his life: the journal entries are punctuated by the deaths of his friends.

Clearly, Jarman had an intensely emotional response to his garden.

Entries often go like this: "I have never been happier than last week. I look up and see the deep azure sea outside my window in the February sun, and today I saw my first bumble bee. Planted lavender and clumps of red hot poker," or "The wind got up, bringing the smell of the sea; a russet kestrel flew by. Extraordinary peacefulness." Gardening bends time for Jarman and takes him out of the day-to-day grind. "The gardener digs in another time, without past or future, beginning or end. A time that does not cleave the day with rush hours, lunch breaks, the last bus home."

He quotes Aldous Huxley on the state of being alone—really, always—but still empathizes with the garden on an interpersonal level ("I could feel the garden sigh with relief") and finds a kinship with the "shiny poppies crumpled like party hats" and even the landscapes which like dying lovers are "shrivelled and parched."

The garden is the centre of surprise, he writes. It is also a place of structure, routine and patterns of expectation. Even if the strong winds kill everything off, there'll still be something jolly. "Bright sunlight with a strong gale blowing. I repaired the garden for the third day running. All the plants dreadfully damaged: the wallflowers look as if they have been boiled; even the lavender has wilted; and the crocuses have shrivelled in the bud . . . Beneath the willows at the Long Pits I found the first primrose of the year."

He doesn't know when he waters the roses if he will see them flower, but watching the other plants spring up gives Jarman hope. Hope, beauty, colour and love don't make fear of death or grief vanish, but they do seem to make life easier. Towards the end, his attachment to his garden—to living things, his plants, his flowers—sometimes makes dying harder. While it can soothe his grief about the death of his friends, the final pages are full of nostalgia and bittersweet longing. "The day of our death is sealed up. I do not wish to die . . . yet. I would love to see my garden through several summers." For Jarman, there are no binary conclusions. It is not that his garden makes dying easier or not. But it does give his life a richness and fullness and a loved character and relationship that is with him until the very end, even in the final loneliness of death.

At the end of the book—when he is in hospital—he frets about his garden. His friends bring him "masses of beautiful flowers" and cigarette cards of country flowers. His mind keeps floating back to Dungeness.

"How I would love to be putting the seed in the garden. It shouldn't be too late if I get it in by April." At one point, he plants the garden in his mind, sowing calendula and fennel. At another, he imagines he's in a bluebell wood, surrounded by splashing streams, huge old trees, moss and wildflowers. He continues to sow poppy seeds until the very end.

PART VI

# SNAG

# 10

# Future Nature

Gardening is the most therapeutic and defiant thing you can do, especially in the city—and you get strawberries.
   —GARDENER RON FINLEY, who plants vegetables on kerbsides, abandoned lots and traffic medians in South Central Los Angeles

The twentieth century has eliminated the terror of the unknown darknesses of nature by devastating nature herself.
   —THOMAS BERRY, "Technology and the Healing of the Earth"

S VALBARD IS AN ARCHIPELAGO of islands, islets and skerries at the entrance of the Arctic Ocean, north of Norway. It's so far north that you have to look south to see the Northern Lights. It is the home of the fictional armoured bears, or *panserbjørne,* in Philip Pullman's *His Dark Materials* trilogy, though of course they couldn't do anything to protect the native flora and fauna from hunters over the centuries. At various times in history, Svalbard was used as a base for Europeans to catch walruses, bowhead whales, Arctic foxes and polar bears. Since 1612, hunters from different areas of the world have sailed to the islands to find rich natural resources. Whalers particularly sought the blubber of the bowhead whale, also called the Greenland right whale, which they boiled into oil on the beaches in large copper cookers. At one time, Predator X, the largest marine reptile from the age of the dinosaurs, swam its waters, all fifteen metres of rapid, prey-catching hulk, longer than a double-decker bus.

Few people currently live in Svalbard, just a couple of thousand. People who ship in don't stay for long, and the mining operations stopped years ago. In winter, the streets are dark at lunchtime and you need a torch to get around. For one month, there isn't even a twilight glow. The Norse word *svalbard* simply means "cold shores." It is a land of ice; quiet and windswept. It is also a land of wonder. The Svalbard moss campion (*Silene acaulis*) is an extreme survivor. It has evolved its own central heating system as a way of enduring the cold and dark; its innards can reach 30°C even if the outside temperature is freezing. As the sun passes from south to north, so the pink flowers on each green dome bloom from south to north. Explorers used the flowers to find their way.

Svalbard is an extreme place for anything to survive. Just 13 percent of Svalbard's land is covered in vegetation. Over half is glacier, and barren stone makes up 27 percent of the landscape. Walruses live in the waters, searching for mussels, snails, crabs and even ringed seal. They can hold their breath for forty minutes and dive down almost half a kilometre, using their whiskers to find mussels, then sucking out the flesh and discarding the shell. On land or ice, they lie close together, even sometimes on top of each other, for they are sociable creatures. In winter, Svalbard reindeer, Arctic foxes and ptarmigan have different ways of protecting themselves from heat loss, though they all increase their bodyweight by almost a third in the autumn. The ptarmigan grows a thicker coat of white feathers for the winter and added feathers around its feet, toasty slipper-socks to see it through the cold months. The Arctic fox rolls itself into a ball and lies in the snow, keeping itself warm by using its bushy tail as a blanket. The reindeer, and the fox, shift from their lighter summer coat to a more insulating winter pelage. Passerine birds—snow bunting and wheatear—are the only birds that travel to Svalbard to breed.

It isn't an easy place for humans to live either. A taxi driver told me the ladies go crazy in Svalbard because they have nothing to do. Then they move. Come here if you want a divorce, he said. The darkness breaks up families; they cannot cope. In the summer, the light drives people to drink. It was too much, he said. There is a price to pay for the scenery, the peace and the light, the psychic demands of living in the Arctic. Residents and visitors must stay within the perimeter of the settlement, for polar bears roam the landscape. To leave, you must either be armed or in a car. Safety advice posters suggest shouting and growling and waving arms, if you see a bear. People have the right to shoot in self-defence, but all shots have to be reported to the governor of Svalbard.

The light in Svalbard is exquisite: both lucid and hazy, reflected by cream and gold mountains. As the sun moves, the mountains glow ultraviolet pink at their tips and the rest is a pale violet, against the bluest sky. Across the strait, a mountain soars up into a flattened peak, crenellated like the spine of a large dinosaur. The mountains are stolid but seem completely alive. Light and shade fall between their wrinkles and folds, which look like children hiding under sheets, as if they could stand up at any time. Later in the day, when the sun begins to creep away, it diffuses a fiery neon pink-orange across the slopes and plains. Then the mountains turn lilac, yellow gold and, finally, a cold blue. Some natural philosophers, according to the eighteenth-century physicist Horace Bénédict de Saussure, believed that the colours rose from "accidental vapours diffused in the air, which communicate their own hues to the shadows."

The houses in Longyearbyen, the only settled town in this vast wilderness and the northernmost town in the world, are painted in bright colours: ochre, olive, jade, maroon, terracotta and mauve. Cheerful, you would imagine, in the face of polar night and polar light. From afar, the dwellings look like a model village, a Christmas toy town in a shop window, with reindeer and parked snowmobiles here and there. Shops sell bright-red fizzy drinks and salty Asian snacks. Smells carry and linger for longer. Wafts of baked goods hang around for twenty steps. The smell of diesel from a car remains for seconds longer than normal, with nothing but sugar-like névé to soak it up. Reindeer wander around, and every few metres a yellow splash of their urine punctuates the white, snowy path. It is a quiet place to explore, the unending space of the landscape creating a silent peace, interrupted by the crunch of ice underfoot or a car engine. The cold is bruising to any part of the skin not covered; it feels as if the blood of my face is hardening, changing consistency into gel.

One of the reasons I was drawn to Svalbard was because it is the largest wilderness in Europe, and one of the last wildernesses on Earth. I wondered if I'd find there, as Christiane Ritter has detailed in her extraordinary account of living in Spitsbergen in 1933, consolation, elevation of the mind, high spirits and serenity. "Nature seems to contain everything that man needs for his equilibrium," she wrote in *A Woman in the Polar Night* (1954).

At the water's edge, it isn't easy to imagine a sea so filled with whales that ships had to break their way through, as was reported by a sailor called Captain Pool in 1612. The water gently undulates, reflecting the pink light of the mountain sun. A mist rises from the water, a strip of lint-like fog. I feel a sense of peace, as if I were suspended. I feel ant-like in the face of the mountains. The currents here are warm, and Svalbard is known as the "tropical" Arctic Archipelago. The Arctic Ocean is biologically diverse and rich, which brings many scientists here to live and study. On the beach, the seals roll over like newborns, this way and that, paws in the air. The sea in Longyearbyen harbour, a local told me, stopped freezing over about ten years ago. The ice in this area of the Arctic has reduced significantly. In satellite images that show the Arctic sea ice extent between 1980 and 2011, it looks as if someone has taken greedy spoonfuls out of an ice cream.

My primary reason for visiting was because it is the home of the Svalbard Global Seed Vault, built in 2008 to store and protect hundreds of thousands of plant seed species in case of global disaster, ecocide, war, famine and the other possible effects of climate change. The vault is a safety net for impending ecological collapse. A solution of sorts (albeit chilling that it is even needed) for the way we live now. It was a symbol, to me, of our future, predicting what might come next. If we follow the trend of ecological collapse to its end point, seed vaults will be an important way of conserving nature, for both our physical and our psychological sustenance.

There are currently hundreds of seed banks across the world, but Svalbard has the largest collection of crop diversity. It houses almost a million samples and there is room for millions more. It doesn't cost anything for countries to store their samples, but they have to make them available for requests from scientists or farmers who want to try to grow a new variety of plant.

The primary reason for freezing the seeds in Svalbard is agricultural, the ultimate back-up for crop safety, in an age where traditional knowledge about farming and crops has diminished. It sounds extreme, and futuristic, but in fact the first withdrawal took place in 2015. Following the ongoing war in Syria, scientists couldn't access the vault in Aleppo, so they retrieved seeds from Svalbard to set up a new facility in Morocco.

If you drive a little way from the settlement you will see, lurching out of the side of one of the mountains, looming over the frozen tundra, a door that resembles the monolith in *2001: A Space Odyssey*. The top section of the door is decorated with shards of triangular mirrored glass which reflect the light. Inside, it is very industrial. At the entrance, bright blue hardhats are piled up on a bench to the left underneath fluorescent strip lighting. A wall of white storage cupboards is on the right, with large metal pipes running along the ceiling.

"We are walking, right now, to the past, present and future of agriculture," said Brian Lainoff, Lead Partnerships Coordinator of the Global Crop Diversity Trust, as we stepped into the mountain. He is a young American Italian and infectiously evangelical about crop diversity. "These are the seeds that will without a doubt go on to feed us, they will go on to create new and better crops that can resist disease and climate change."

We walked through hollowed-out rooms and corrugated chambers, making our way to the heart of the vault. Boxes of seeds—wheat, pea, pumpkin, niger, potatoes, rice—were waiting to be taken into the vault itself. Each wall and door was crystallized with frost, some of which feathered and filigreed into beautiful geometric patterns. We jumped in and out of the vestibules as it became colder and colder, to avoid letting in warm air. The door of the main room was encrusted with glittering ice, almost covering where it started and the wall ended.

Brian opened the door and, strangely for an apocalyptic stronghold, it was rather small and basic. Shelves of plastic boxes, marked in primary colours: a pantry-sized IKEA warehouse. North Korea sat next to the USA, South Korea next to Japan, Ireland next to Canada. It was moving to see almost every country in the world safeguarding their crops for the "common good."

In May 2017, however, the seed vault in Svalbard flooded—after a period of melting ice and heavy rain—to the Crop Trust's surprise. The seeds weren't touched, but it raised fears that as climate change worsens the vault might not be as ecocide-proof as it was thought to be. The very meaning of "permafrost" is falling apart in our changing climate. Longyearbyen is now the fastest-warming town on Earth, warming three times faster than the global average because of its location in the Arctic.

To buffer the destruction of nature, many other seed banks are racing to protect their native plant life—and not just for agricultural purposes. Kew, the botanical garden in south-west London, runs the Millennium Seed Bank at Wakehurst, West Sussex, which holds seeds of wild UK species as well as crops. It currently houses almost all species of UK native flora and is collecting seeds of three thousand species of trees. In 2013, the UK National Tree Seed Project was launched, in response to the threat to trees from land use, development, pests and disease. By April 2018, ten million seeds from sixty species had been collected.

We are entering an uncertain period. In light of the fact that billions of trees are cut down each year, and destruction of forests is increasing, with Brazilian president Jair Bolsonaro changing the laws that protect the Amazon rainforest and leaving it open to further deforestation, estimates that there will be no more trees in three hundred years don't seem so far-fetched. Nor does the Svalbard Seed Vault seem like a storyline from an episode of *Doctor Who*.

ALONG THE STREET where I live, there is a large, white, listed Art Deco building which was built in the 1930s, around the same time as my house. It's empty and disused, but inhabited by seagulls, pigeons and, I reckon, nesting kites. It looms large on the skyline—smart white chunks, as if it were made from Lego. Soon, it will become luxury flats. Before it was abandoned, it was Eli Lilly's main factory in Britain for the production of various medicines. From the 1980s, capsules and oral solution of Prozac were manufactured in the building and sent out to chemists and hospitals across the country. If I sit in a particular place in my garden, I can see it and imagine it filling up with the millions of small green and white pills that were produced there, to alleviate the depressive symptoms of thousands of people who needed psychic relief.

Personally, I am hugely grateful to antidepressants. I've taken one sort or another for most of the last decade. The abandoned Prozac building, however, is a symbol to me of the limitations of holding up psychiatric drugs as a universal panacea for everyone suffering from depression. They can't be. Nor, in Britain, can the financially strained NHS currently offer the kind of long-term, face-to-face psychotherapeutic help that everyone with a mental health problem might benefit from at the point of crisis.

But if we shifted towards a different model of healthcare, one that considered the psychological and social factors behind health that we now know play a large part in various diseases and disorders, we would broaden the scope of both preventative measures and treatments. Precision medicine is being hailed as the next major breakthrough in healthcare, with both President Obama's 2015 Precision Medicine Initiative and the NHS in Britain "embedding a personalized medicine approach into mainstream healthcare." With data-driven treatment, and the technology that will make treatment more precise (patient-generated data through smartphones and wearable tech, genomic medicine and computer science), we are at a point when it should be possible to break the Cartesian dualism that has historically separated the mind from the body in how we think about health and illness. Without seeing the full picture, we have a much narrower, more simplistic and less effective means of treating and preventing potential problems.

Connection with the natural world has numerous psychological and physical benefits far beyond what our society gives it credit for. Without opportunities to commune with greenery, other species, trees, the stars, we are poorer, sicker and more stressed-out. Contact with the rest of nature increases capacity to cope with stress, enhances life satisfaction and general outlook, helps recovery from mental fatigue, restores concentration, aids recuperation and improves productivity. Though, of course, it is not a panacea itself.

What does the future of our psychic relationship with nature look like? In the modern world, much of our contact with nature is through technology, and evidence suggests accessing nature in this way (see the Snake River project in Oregon, described in Chapter 6) could have some benefits. Much of our exposure to the natural world is through nature documentaries—and they are enormously popular. In 2017, the most watched TV show in Britain was *Blue Planet II,* the follow-up to David Attenborough's landmark series about the ocean. In China, it was downloaded so many times it slowed down the country's internet. Engagement on social media about nature programmes brings people together (as anyone who follows the BBC franchise *Springwatch* on Twitter will know).

A variety of apps for smartphones allow people to identify trees, birdsong, constellations and, in the case of Schmap, mark the things they enjoy seeing. On Instagram in the springtime, photos of blossom fill the platform's squares. When I check my feed, I will always see at least one image from the natural world, be it sunsets, macro photographs of pollen seeds, wildflowers, elaborate houseplants or insects. It is technonature connection. And it might be all we are left with.

Studies have shown that just looking at a picture of the natural world can give stress-reduction and well-being benefits, but is it as powerful as the real thing? Could technonature replace our need for real nature? Can it give us the same wide range of psychological effects? Is a tweet of emojis of a woodland enough? Will PARO seals, the therapeutic robots designed to calm people with dementia, satisfy our need for company as we continue to live longer? Peter H. Kahn Jr. is a leading expert in technological nature and well-being. He's compared the effects of looking at a technological nature window compared with an actual window view. The participants were measured for physiological recovery from low-level stress. Those who looked through the glass window at real nature saw a quicker decrease in heart rate than those who looked at the technological nature window. The results suggested that a technological nature window is better than nothing, but it's not as good as the real thing. I order a VR headset to give it a go. I watch sharks sweeping the ocean floor, lions lounging, Saturn. It hurts my nose where it rests. I watch videos that are being trialled for people with chronic illness, in healthcare settings. Certainly it is preferable to looking at the mess of my living room, but I feel a little sick afterwards, with a slight headache. VR nature has its place. If I was housebound I would watch it daily. But I think that if this was it, my heart would break.

In 1995, the Telegarden was launched by robotics artist Ken Goldberg, of the University of Southern California, and Joseph Santarromana, of the University of California, Irvine. It was a community garden tended to by internet users controlling a robotic arm. They planted seeds, watered plants and talked about their gardening in an online chat room. Over ten thousand people logged in as gardeners over the decade it was online. Kahn examined patterns of conversation and discussion to deduce how engaged people were with the garden and whether they talked more about nature in the setting, or outside or even at all. Telerobotics could be an interesting way of connecting people to the natural world, but would it fall short of the real thing?

Again, the results were not promising. Only 13 percent of the conversations were even about the garden or a relationship with nature outside the interface. The conversation within that seemed "thin." "People did not talk about the plants in biocentric terms or deserving care and or respect," wrote Kahn. Examples included: "my watering is done" and "Hold on a sec, let me plant one real quick." In conclusion, said Kahn, the Telegarden interactions "did not seem to satisfy the mind, heart and spirit of what most gardeners presumably experience." The artists who designed the Telegarden conceded that they were trying to make a point about the internet, in its earliest days. "Maybe it's time to get off the internet and out into the garden," said Goldberg.

And what if that garden is made of fake lawn and fake trees with fake houseplants inside? As current trends continue, and we eventually singe the planet until only the cockroaches survive, could we simulate a natural environment using plastic or man-made materials? Or, if parks seem too expensive to local governments, could we just replace them with Astroturf and pipe out birdsong, and reap the same psychological benefits?

For a start, such a canvas would mean no insects, birds, animals or microorganisms, no oxygen, no sequestering of carbon, no handy microbacteria. It would mean no smell or taste, unless someone could invent nice-smelling plastic. It would be a much more impoverished sensory experience. The heat-reducing properties of trees and grass in urban areas would be removed and the danger of urban heat spots increased, as Astroturf can make the atmosphere hotter. Personally, I don't think I'd like to lie on plastic grass.

This isn't some crazy dystopian nightmare. Lots of people are covering their lawns with Astroturf now because it's easier not to have to mow it. In the supermarket the other day, the man serving me at the till—a keen gardener, I was buying compost—lamented the fact that his son was covering his garden with Astroturf. "I don't understand it," he said. "What about the butterflies and the bees?" The one benefit you can get is the restorative properties of green, but compared to real nature, that's fairly meagre. We are becoming so divorced from nature that we don't even have real gardens any more.

Some people say we shouldn't worry: we will adapt to the loss of nature, as we are an adaptive, technological species. Kahn's response is simple: "We *are* an adaptive species. But not all adaptations are good for us." We don't know exactly where technonature will go in the future. Nature documentaries have improved immeasurably over the decades, and perhaps telegardens and technological nature windows will do the same. It is likely the field will become more advanced and sophisticated and offer greater psychological benefits. But I am willing to bet that technonature will never match what is already there or its impact on our nervous systems, neural pathways, imaginations, existential angst, gut, stress resilience, memory, ability to experience awe, peace, joy and tranquillity.

AS I'VE MENTIONED, by 2050, 68 percent of the world's population will live in urban areas, and face the stress that goes with it. As of 2018, in Northern America 82 percent of people already live in urban areas; in Europe it's 74 percent. If the current extinction predictions are correct, and the rate of habitat destruction continues, the opportunities to commune with the rest of nature will be significantly diminished by 2050 for everyone, but particularly for those who live in cities and towns. Biodiversity loss will deplete the natural environments where some humans go to seek restoration, and thus the extent and quality of that restorative effect will shrink. But cities aren't going anywhere, and people need homes, so how can we make modern life a happier, greener place for all people?

The biophilic city movement offers a vision of reshaping the primary human habitat that allows for nature, and incorporates the non-human world into all aspects. Another phrase for it is "green urbanism," which imagines cities in a new, healthier, biocentric way. Does it ever irritate you when you're in a town and it takes ages to cross the road because you have to wait for car after car after car, breathing in the fumes, waiting until they all go by? Or, if you're a cyclist, when you can't cycle somewhere because there aren't any cycle lanes and it's impossible to navigate through roundabouts and highways and dangerous lanes? Green urbanism and biophilic city design calls for walkable neighbourhoods and bicycle-friendly towns, alongside sustainable public transport.

Another element of biophilic cities is the integration of growing food into the way the city is designed, in order to improve health, biodiversity and food security. Most towns and cities already have a certain number of trees and gardens and plants, but in a future biophilic city it simply wouldn't be an option to cut down thousands of trees in, for example, a city like Sheffield, nor would they be a tick-box afterthought. Political endorsement will be vital.

DETROIT IS A NOW biophilic city—though perhaps an unlikely one, because for decades it was the behemoth of the US automobile industry, in the early to mid-twentieth century. But by the 2000s it had become the poster-child for urban decay after decades of disinvestment: tens of thousands of buildings left empty, burned down, collapsed or in a dilapidated state; people dying young; the unemployed forgotten. A consequence of the open land and vacant lots was a unique landscape, and it wasn't left empty for long. Members of a burgeoning grassroots movement realized there was an opportunity. They started planting and transformed the city by bringing nature back in.

Now, Detroit has more than 1,500 community gardens and small urban farms of two to three acres driven by the community's desire for real, fresh, nutritious food after retailers left the decaying city. Fresh food is crucial for the population of Detroit, who suffer health problems caused by poverty, lack of healthy food and historical systemic racism. The urban agricultural movement is a defiant act against the situation Detroit found itself in, following racial violence and the middle-class flight to the suburbs, leaving the population "for dead," as Devita Davison, Executive Director of FoodLab Detroit, has described it. People of colour, historically excluded from agriculture, owning land, access to resources, access to the natural world, access to nutritious food, are reclaiming Detroit by planting and growing.

As well as the nutritious value of growing collards, cabbage, broccoli, kale and other vegetables, the community aspect is central. People grow together; they collectively buy abandoned buildings and turn them into kitchens or cafés. "These are spaces of conviviality; these spaces are places where we're building social cohesion as well as providing healthy, fresh food for our friends and family and neighbours," Davison said. "Every day, we collectively work to redesign the whole food system, from farm to table, while honoring and creating space for people of color at each step . . . We're creating a new society. They're stories about love, for community, each other and Mother Earth."

Almost ten thousand miles away is another, very different biophilic city-state with a much higher GDP and level of infrastructure: Singapore. The country is leading the way with green rooftops, green walls, green balconies and vertical gardens. The Khoo Teck Puat Hospital, which was opened in 2010, is particularly radical, integrating "forest-like" nature with an understanding that while their patients and staff need it, they also need to conserve and protect species (butterfly species living in the grounds of the hospital have grown from three to eighty-three). It was designed by the ex-CEO of the hospital so that "one's blood pressure lowers when he/she enters the hospital grounds." Anyone can visit the hospital and see its species of butterflies and birds, flowers and plants, insects and foliage. For patients, there are balconies filled with scented flowers, plenty of natural light, natural breezes that carry over Yishun Pond, organic food made from vegetables and herbs grown on-site. Even though it is in a dense urban area, it has a green plot ratio (an ecological measure for architecture and planning) of 3:92, which means the surface area of horizontal and vertical greenery is almost four times the size of the actual plot of land the building sits on. In a study of patients, staff and visitors, its biophilic design elements were cited as the reason it was preferred to an older, more traditional hospital without nature. Of the two hundred surveyed, 80 percent were in favour of widespread investment in bringing nature into hospitals. It also consistently beats other hospitals in Ministry of Health public satisfaction surveys. In 2017, it won the Stephen R. Kellert Biophilic Design Award.

Humans have long used nature as an aid to recovery and rehabilitation. We have known empirically, since Roger Ulrich's landmark study (see Chapter 2), that people recover more quickly from surgery in hospitals with views of a natural setting. The Khoo Teck Puat Hospital offers a blueprint for how our hospitals could look—for the benefit of the unwell, the wider community and the species that live in it.

In Europe, Milan's Bosco Verticale (Vertical Forest) is the ultimate green skyscraper and, the architects hope, a model for a sustainable, residential building in a built-up area that enhances urban biodiversity and nature contact. The balconies are planted with 4,300 bushes, 15,000 plants, shrubs and flowering plants, and 800 trees, including beeches, oaks, yellow maples, ash and yellow acacias, which grow up to nine metres high.

The rest of nature should be incorporated—with urgency—into the way we design our cities and homes, even the areas we tend to overlook. Lauriane Suyin Chalmin-Pui, a University of Sheffield PhD student, is currently researching the therapeutic benefits of front gardens, an "endangered species," and how people engage with planted-up front gardens compared with paved-over parking spots or empty lots. As insect decline goes into free-fall, we should use empty land to plant flowers for pollinators and other organisms that keep us alive.

And perhaps—whisper it—we need to rethink our lawns. In the United States, there are forty million acres of lawn, swathes of monoculture. Grass is in fact North America's biggest crop. But grass supports little biodiversity—especially if, as is becoming more common in drought-affected areas such as California, the grass is sprayed with green paint—and we are in a biodiversity crisis. Is it time to reassess our beloved traditional lawns and consider how we could better use the space? United Kingdom–based scientist Lionel Smith has been researching how to create grass-free lawns and suggests red flowered daisies (*bellis*), white flowered buttercups (*ranunculus*) and bronze-leaved bugle (*ajuga*) as examples of plants that would work in the climate of Western Europe.

Biophilic design is in its early days. In Britain, we have few biophilic buildings; the Barbican Estate is perhaps the most famous. Built on a large area that was bombed in the Second World War, the estate was built between 1965 and 1976 around a lake, with forest trees, lawns, waterfalls and now palm trees, cacti and colourful roof gardens. It seems strange, considering how celebrated it is, that it remains one of a kind.

In the United States, in Milwaukee and San Francisco, people are planting up car parks and growing community orchards. In Australia, Melbourne is aiming to double its tree canopy coverage by 2040 as part of its "Urban Forest Strategy" led by the City of Melbourne's Urban Sustainability Branch. The City's government has recognized that an urban forest will "play a critical role in maintaining the health and liveability of Melbourne," a city which is threatened by climate chaos and extreme urban heat island effect.

Protecting the sound and song of the natural world and controlling man-made noise is also a consideration of biophilic designers. Currently there is a protection order on silence in the Hoh Rain Forest area of Olympic National Park in the State of Washington, one of the most ecologically diverse areas of the United States. Acoustic ecologist Gordon Hempton runs the One Square Inch of Silence project which protects the area from human noise by, for example, asking airlines to reroute flight paths. Listening to a recording online, of course it isn't silent at all, although it feels intimate. You can hear the rain falling, droplets hitting and sliding from the leaves and branches of the moss-covered maples and enormous spruces and into the soft undergrowth. Birds patter and cackle and trill. It is easy to imagine the mist rising and the smell of the pines from a thousand miles away. It isn't just human mental health that will benefit from the preservation of natural sounds: wildlife relies on a sense of hearing to pick up mating calls or sounds of prey. As more of us move to urban areas with high levels of background ambient noise, protecting small pockets of peace and natural quiet is essential.

There are many unarguably good reasons for biophilic cities. Greening the land leads to community, which fosters connection and, as we have heard, greater interest in other people. Neighbourhoods would be protected from stressful noise with peaceful areas provided for citizens. In an uncertain changing climate, societies need enhanced resilience and social connection—and a biophilic culture, its proponents argue persuasively, will foster this. But how could it be politically enacted? How, for example, could we get to a world where rivers are given solid protection from environmental waste, and bumblebees are given homes that aren't made of Astroturf plastic? What will it take to reorder our world to respect the wild, and the species that live in it, both human and non-human?

SINGAPORE, home to the radical Khoo Teck Puat hospital, is currently striving to be the world's greenest city. According to research by teams at the Massachusetts Institute of Technology (MIT) and the World Economic Forum (WEF), Singapore has the highest percentage of urban tree canopy in the world. Almost a third—29.3 percent—of its area is covered with green. London, in comparison, has trees covering 12.7 percent and Paris, 8.8 percent.

The government of Singapore has stated that they want to turn Singapore into a "city in the garden." To achieve this, policies and rules have been put in place, particularly the Landscape Replacement scheme. Since 2009, property developers have been required to replace any greenery lost from a site due to development with greenery in other areas within the development—after the building work has finished, and usually in the form of vertical gardens and rooftop greenery. The Landscaping for Urban Spaces and High-Rises programme (neatly abbreviated to LUSH) extended the policy to cover more areas and development types in 2014.

This kind of policy and legislation will be key. And what would it take to protect natural spaces on the basis of human health?

The main idea of Wild Law—or Earth Jurisprudence—is that all components of the rest of nature, including plants, animals, rivers, and even entire species or ecosystems, should be granted legal personality in the same way as human beings. Britain and the United States have a sorrowful history of exploiting the land, and the people who live in it. In our legal system, public limited companies (PLCs) have more legal rights than the natural world. There are currently no environmental laws that protect natural areas for public health.

This sounds a bit, well, wild, but it's a developing area of law with precedents already established in other countries. In Britain, there is a Wild Law special interest group in UKELA, the leading environmental law association. At the end of 2017, Scottish lawyer Colin Robertson suggested that Ben Nevis, the highest mountain in the British Isles, be given personality in law, though this hasn't happened yet.

It's nothing new in some indigenous communities. When a group of Oneida people, a Native American tribe who lived in the area that's now central New York, needed to move to a new territory, there was an obstacle standing in the way: wolves. What would they do about the wolves that already lived in the land that they wanted to inhabit? At the Oneida council meeting, to make sure the needs of the wolves weren't forgotten, a member of the tribe would advocate for the animals. At the beginning of the council, they would ask, "Who speaks for Wolf?"

Internationally, the UN is starting to turn away from an anthropocentric ideology. In the General Assembly Resolution 70/208 of 22 December 2015, it formally recognized the principles of Earth Jurisprudence. It recommended that governance systems "recognize the fundamental legal rights of ecosystems and species to exist, thrive and regenerate," and observed that "they are not in opposition to human rights: as part of Nature, our rights are derived from those same rights." Human rights "are meaningless if the ecosystems that sustain us do not have the legal right to exist," concluded the experts' report.

A number of countries have now changed the way their laws value and protect natural spaces. In 2008, Ecuador adopted a new Constitution which includes the Rights of Nature. In 2010, Bolivia passed a "Law of the Rights of Mother Earth," which established Mother Earth as a "collective subject of public interest" and a legal personality. There is now an ombudsman for Mother Earth.

In 2014, New Zealand passed the Te Urewera Act, which granted the 821-square-mile forest, an ancient homeland sacred to the Tūhoe people, an indigenous group of the Māori, legal rights. The landscape owns itself now, as it were, and the revolutionary legislative change has been heralded as a "new dawn in conservation management." "The settlement is a profound alternative to the human presumption of sovereignty over the natural world," said Dr. Pita Sharples, the Minister of Māori Affairs. This means that the land has "all the rights, powers, duties, and liabilities of a legal person," so that lawsuits can be brought on behalf of the land itself. A new Te Urewera Board was set up with an equal number of Tūhoe and Crown representatives. The 2018/2019 Annual Plan included pledges to reduce pressures on Te Urewera to "enable a rebalancing of her natural state" and to reconnect people with "the healing life potency of Te Urewera." Hunting restrictions have been put in place to safeguard biodiversity, for example. The Whanganui River in the North Island of New Zealand is also recognized as a person in law.

In the Republic of Benin, the Benin Forest Law recognizes the sacred forests and natural sites where gods, spirits and ancestors reside. In India, the heavily polluted Ganges and its tributary, the Yamuna, were granted legal personhood in March 2017: The High Court of Uttarakhand's declaration meant that polluting or harming the rivers or natural areas would carry the same consequences in a court of law as harming a person, which conservationists hoped would be a strong deterrent. However, the order was overruled by the Supreme Court, which argued that the declaration was legally impractical.

Veneta Cooney heads the Public Health and Environmental Law Working Party for the United Kingdom Environmental Law Association (UKELA) and works as a consultant physician in London. She is particularly interested in the link between public health and environmental protection and gave me an insight into how it currently works and how it could change in the future. I enjoyed talking to her very much; she is highly intelligent, warm and charismatic, and joined the dots for me between health, law and the environment in a way that left me feeling excited and hopeful. "When you look at environmental legislation as a whole, the main caveat that gives people the excuse to concrete a garden is this concept of overriding public interest," she says. "This runs through legislation like DNA. Let's protect that park, or that mountain, but only if it's in the public interest. It's this concept of public interest we need to change. To date, it is all about economic growth. We can concrete over a park and put up an office block, because it will create employment."

And how does that affect public mental health? "No one has really thought about the secondary and tertiary consequences of that economic model," Cooney continues. "If you concrete over that park that used to be used by old people and children, where people can be mindful and calm, the mental health consequences of living in a concrete jungle are much greater than what an office block is bringing in."

Finally, the quantitative data has joined the qualitative evidence to prove the association between nature and human health and well-being—and now we need to move fast. "The medical profession who are extremely well placed have been suspiciously quiet," she goes on. "I think that's a shame. For fear of being seen as political, they haven't been vociferous enough."

From Cooney's perspective, British industry and society are still mired in the 1950s discourse of building roads and motorways, and obsessed with progress, development and growth, in a culture where it is normal to concrete over green spaces that remain while ignoring the benefits of nature contact on people with dementia, obesity, loneliness and mental illness. A new holistic mindset is needed, she argues, made acute by the state of the NHS. "The NHS is on its knees, and one of the reasons why is that too many people are sick."

Weaving the concept of mental well-being into new housing developments isn't quite happening yet. After speaking to Cooney, I read policy papers and frameworks published by the UK government and industry guidance on optimum housing developments. There was lots of information about recycling and car park space, but scant mention of mental health, nor its relationship to connecting with the rest of nature. In one document, *Building for Life 12,* published in 2015 and based on the National Planning Policy Framework, wildlife habitats, trees and plants were barely mentioned. It shows just how much our plans and ideas of how we live in the modern world need to evolve.

As I have suggested in this chapter, seed banks, technonature, Wild Law, biophilic cities and a more holistic model of healthcare could all be important tools for future generations to protect, conserve and relate to the natural world. However, seed vaults can be flooded and a technological garden might malfunction, so we can't ultimately put our hope and faith in them. Instead, changing the way we design cities, how we protect other species in legislation and how we think about our health in relation to the wider environment will bring deeper, long-lasting change. Allowing the non-human world to exist on its own terms; treating other species, trees and land as equals, instead of underlings, must be the way forward. But this will only work if we fall back in love with the natural world and form a connection.

# Conclusion

## A New Dyad

We see ourselves as part of the natural world. Everything that moves has a spirit. We are no more important than the wind, sky, grandfather rock, grandmother earth, plants that grow, water that flows, or any of the winged, two-legged, or four-legged beasts. We are all stardust. There's no hierarchy. Everything that powers the universe goes in a circle. This is the sacred hoop. We are all relatives.

—RUTH HOPKINS, a member of the Sisseton, Dakota and Lakota Sioux First Nation tribe

A garden may be a lovesome thing, and a green plant on a windowsill some refreshment to the office worker or housewife; but it is the big tracts of unspoilt natural landscape, plants, animals and all, that are most likely to refresh our tiring spirits and keep us sane.

—ANTHONY HUXLEY, *Plant and Planet*

A T THE MOMENT, the relationship between us and the natural world feels like that of a truculent teenage daughter and her mother: resentment, a lack of attention, careless comments, forgetting or not realizing what the mother has done for her. It is not the teenage daughter's fault: there is a complex alchemy of things happening all at once, from the chemical changes in her adolescent brain to social pressures, both to break away from the parent and also society's perception of motherhood. Of course she still needs her mother, but she doesn't quite realize it, and spurns and scorns her in the bid for independence and freedom and desire and the pleasures of consumerism in the heady context of the Great Acceleration of the twentieth century. Mother is a cringe, an embarrassment, too earnest, too much. The disconnection doesn't help us—it makes us sick and ill and sad—but we can't help but rail against her. A mindless trashing with hurtful words and slammed doors and a distinct lack of harmony. This period doesn't last; it can't last. The daughter still yearns for the comfort of the mother, and the mother's bond to her child will never be broken. But the relationship needs to be restored and recalibrated. We need to remember where we came from, who nurtures and succours us, and how much bounty Mother Nature offers us. At some point, the daughter realizes she needs the mother again. And she understands what the mother has given her.

How can we collectively fall in love again with nature? How can we think about the relationship differently? By knowing and noticing the other beings who live around us, for a start. In the same way that agony aunts recommend to couples whose relationship may have become stale that they should remember what it was that attracted them to their partner in the first place, in order to rekindle the love, we need to remember that childlike wonder at nature as adults. Take a step back, and step in to reconnect and to wonder and to realize the intrinsic interconnectedness of Planet Earth.

I started investigating the mechanism by which contact with the natural world might affect human mental health in 2012. In the last eight years I have been surprised by how much and how varied the evidence is. Starting out, I thought there might be a silver-bullet study or piece of work that explained why natural environments are mysteriously restorative. Instead, it is a salmagundi, layer upon layer of positive benefits to different parts of the body, brain, mind and spirit.

Beforehand, I might have said that a relationship or connection with the rest of nature isn't for everyone; that, of course, some people don't like the outdoors. But in fact the research shows that background nature is essential across the population for good mental health. Alienation from the natural world is eroding our mental health, even if identifying insects or climbing trees isn't how you'd choose to spend your free time. Without access to natural landscapes, the extent of biodiversity, flowers, plants, animals and trees, our potential for restoration, peace and psychological nourishment is sorely degraded.

I presumed initially that nature-based therapy wouldn't be effective as treatment for the more severe mental illnesses; that it was helpful for restoration and offering a remedy from normal cognitive fatigue, but that it was unlikely to help people with schizophrenia or suicidal thoughts. That idea was quickly proved wrong.

My research has convinced me that alienation from the natural world is a factor in the mental health crisis in the West. If we are disconnected from the natural world, we are missing out on nourishment for our minds. We are living in cosmic and social exile and in isolation from other species and elements we evolved alongside. We would be happier and healthier with a richer, fuller, less destructive relationship with the rest of nature.

It can get worse or it—we—can get better. We can stop and look to the future and see where we are heading and how it's going to feel. Or we can continue as if nothing is happening as the planet and its variety of life deteriorates.

Surely, our attitude and mentality towards the natural world and our environment has to change course. The trajectory of destruction has only one end—and as we can't photosynthesize, we cannot "spin the fabric of living matter" if we have killed everything. We must move away from an extractive mindset, towards kindness, care and respect for other species. Our leaders and politicians must move away from short-term, inertial thinking, resisting the radical change we need and underplaying the scale of the crisis, for it is corrosive. The relationship between people and nature is so deeply damaged, and the consequences of our disconnection are becoming so much worse, that a society-wide emergency turning is needed. We must realize the magnitude and seek a mutualistic relationship.

The Hero myth—the drive to seek safety, control and power over the Earth—that has powered Western capitalism and civilization has gone too far. We have taken too much and set patterns in process which are winding up in disaster. The human desire for things, comforts and luxuries has competed with nature to the breaking point. Living to excess must be replaced with living responsibly and sustainably.

How do we as citizens put things right? By, as Jung wrote, "touching nature from the inside." Walking in the woods, lying on the grass, swimming in the sea and practising gratitude for our food, the air we breathe, the water we bathe in, the animals and stars and sunsets we delight in.

Seeking the "numinous" again, the sacred and the awesome in the inter-connectedness of the shrub outside your window which is teeming with life and interactions and brio. By teaching the natural world not as an afterthought; designing our cities with other species as a priority; ensuring that the vulnerable and marginalized have access to natural spaces. Improving and investing in environmental education. If we are lucky enough to have a garden or a balcony, planting flowers and plants that are beneficial for wildlife and particularly pollinators. Not keeping our gardens too tidy, and allowing bugs and worms and bees and wasps to make their homes alongside us. Protecting urban forests from being depleted into less diverse parks and lawns. Culturally stigmatizing Astroturf, which has no place in a dying world. Aligning our economic and political systems with our environment. Employing biophilic officers in local government, who would organize seed collecting in order to enhance and improve habitats for the insects we need to keep us alive. Banning the netting of trees and hedgerows by housing developers to prevent birds from nesting. Issuing wildlife identification cards in different languages to help families learn about nature. Taking a long, hard look at what we need, and what we don't need. Taking only what we need and giving back in return. Keeping fossil fuels in the ground. Looking to traditional ecological communities who have smarter ways of keeping ecosystems healthy by not taking too much from the earth and maintaining a balance. Recognizing that our greatest gifts are outdoors, the very basic materials that keep us alive. Integrating the rest of nature into public health. Supporting farmers and food producers who nurture the land; questioning practices that don't. Rethinking how we use the land to produce food. "Becoming indigenous to place," as Robin Wall Kimmerer puts it. It will require unprecedented change, and time is not on our side.

We don't want to live in a world without bumblebees, whales, conkers or hedgehogs, but really we already do. The idea of it is abhorrent, but it is here, and it could get a lot worse. So it is time for a new story, a new myth, a change of mindset, attitude and behaviour. If we feel it, we must be galvanized by our ecological grief.

It is autumn now and the trees are blazing. I am walking down my favourite local patch, an abandoned canal outside Basingstoke. I am alone but not technically. There is an eight-week-old foetus in my womb. My body is growing kidneys, arterial chambers and either ovaries or testes this week. When I started writing this book, I felt stricken with guilt and worry about the world I had brought a child into. The future seemed frightening and I feared that in a few decades she would have to cope with climate chaos that would cause apocalyptic global instability, on a planet that scientists said could become uninhabitable during her lifetime. Yet part of me wanted her to have a sibling so she would have someone to weather it with, despite my guilt and conflict about bringing another little carbon footprint into the world. Now, though, today, I am more hopeful. In the late 2010s, there has been a shift in awareness about the environment and how we are affecting it (plastics, extinction, emissions), and a similar shift in attitudes to mental health, dismantling the taboos and looking at health in a more holistic way.

Children are mobilizing against climate paralysis with school strikes. Movements such as Extinction Rebellion and Sunrise Movement are gaining speed and traction. The reintroduction of species in Britain has had successes with beavers, water voles, ladybird spiders, bitterns, avocets and otters. There is a renewed thirst for the wild. Publishing has seen a great boom in books about recovery through nature, from bee-keeping to wild swimming, river boating to bird-watching. Between 2010 and 2016, the proportion of people in Britain who visited a natural space increased from 54 percent to 58 percent. It's a small increase, but I wonder if there is something fundamental about our modern techno-society—virtual relationships, social media which can lead to introspection and negative thought processes—that is crying out for rewilding and opportunities to look outward and see ourselves in the bigger ecosystem, as a species among other species.

We can no longer ignore the climate scientists and ecologists who say that time is running out. I am hopeful that a new relationship with the Earth is forming, one which positions us not as conquerors, but co-tenants with wildlife and rivers and mountains and trees, respecting and caring for natural spaces because it is the right thing to do—and because we need the rest of nature both for our lives and for our sanity.

# Epilogue

Granny stepped onto her balcony to see if she could spot Xena walking down the street. It was a warm day but the plane trees provided shade and relief, and Granny brightened at the acid greenness of the fresh spring leaves. The balcony was carpeted with clover and Corsican mint and it smelled good as she walked over it, releasing the fragrance into the air. The clematis and honeysuckle flowers that grew up the walls of the apartment building gently dipped as bees landed here and there. Pink, purple, violet, pale cream, gold. Butterflies powdered the air. She saw her granddaughter turn the corner and watched as she walked on the pavement in her sandals and sunhat. Xena stopped, and bent down next to the central pond, in deep concentration for a couple of minutes. What was she looking at? Frogspawn, perhaps.

Granny returned inside to prepare a snack for Xena, who was always hungry. The first strawberries of the season had arrived in the block's kitchen garden, and Granny had bought a basket for a special treat. She laid the shiny red fruits out on a plate and paused to smell their sweetness. She walked to the other side of the room to look out across the urban wetlands. She picked up her binoculars and spotted a redwing by the rust of his breast. Then, suddenly, she heard it. The screaming. High-pitched and pulsing. Were they back? She looked up to the sky and saw one, then two, then three swifts wheel into view, swooping down to the water and rolling up again into the sky, as if they were on a roller coaster.

The doorbell rang and Granny pulled herself away.

"Hi, Granny, there's frogspawn in the pond!" said Xena.

Granny loved hearing her tales of what she'd seen; it had been a long time since she'd been well enough to leave the flat on a regular basis. "Wonderful," she said. "Come and look at this."

"Swifts!" said Xena, as Granny passed her the binoculars.

They sat for a while together, listening to the drone of the bees, watching the darts and the droops of the swifts, eating strawberries, smelling the honeysuckle.

# Acknowledgements

With very special thanks to Victoria Wilson, my editor and publisher. To be published by Pantheon is a dream moment in my writing life. Thank you, with immense gratitude, to Jessica Woollard, an agent extraordinaire. Thank you to the superb team at Pantheon: Lawrence Krauser, Nicole Pedersen, Cassandra Pappas, Altie Karper, Michiko Clark, Rose Cronin-Jackman, Morgan Fenton and Marc Jaffee. I'm very grateful to Becky Comber for the exquisite jacket art and Jenny Carrow for the jacket design. I was fortunate to have not one but two terrific editors for the UK edition. My thanks to Helen Conford, whose astute eye and passionate nous shaped *Losing Eden* into the book it needed to be, and also to Chloe Currens for her brilliance in the final stages.

Thank you to the Authors' Foundation and the Society of Authors' K Blundell Trust Awards for a grant which made the book possible. Thank you to all the scientists, academics, doctors, nature-lovers and therapists who generously gave me their time. Thank you to early readers and reading recommenders: Panda Gavin, Naomi Escott, Ed Harkness, Sophie Mason, Antonia Peck, Heather Joy, Suze Olbrich and Duncan Cowan Gray. Thank you to Tomasz Samojlik. Thank you to Kara, Brooke, Katie and Bill.

My family has been a crucial support. Thank you, Mum, for the writing weeks in Scotland, and colour, light and landscape. Thank you, Dad, for birds, walks and pygmy shrews. Thank you, especially, to Brenda, Diarmuid and Marlene, for looking after E, and Ed, Ellie, Lexi, Lisa,

Fiona and the rest of our village. Thank you to my grandmothers and late grandfathers for seasides, shrimping, butterflies and peacocks. Thank you in particular to Granny Sinclair, for use of the "apple store."

Thank you, hugely, to Jim, and my little Lark and now Swift, for inspiration.

# Notes

### INTRODUCTION · THE BABY IN THE SOIL

4 Swifts: The British population of swifts declined by 51 percent between 1995 and 2015, and the rate of decline is speeding up. It was down 24 percent between 2010 and 2015. See D. Massimino et al., *BirdTrends 2017: Trends in Numbers, Breeding Success and Survival for UK Breeding Birds. Research Report,* British Trust for Ornithology, 2017, https://app.bto.org/birdtrends /species.jsp?s=swift&year=2017.

4 swallows: Swallows across Europe have been in widespread decline since 1970. See RSPB, "Why swallow populations fluctuate," https://www.rspb.org.uk /birds-and-wildlife/wildlife-guides/bird-a-z/swallow/population-trends/.

4 hedgehogs: D. Carrington, "Hedgehog numbers plummet by half in UK countryside since 2000," *The Guardian,* 7 February 2018, https://www.the guardian.com/environment/2018/feb/07/hedgehog-numbers-plummet-by -half-in-uk-countryside-since-2000.

4 ancient woodlands: Over the twentieth century and early twenty-first century, half of all woods in the United Kingdom that are more than four hundred years old have been lost, along with the decline of species who live in wood- lands. J. Vidal, "UK's ancient woodland being lost 'faster than Amazon,'" *The Guardian,* 21 October 2008, https://www.theguardian.com/environment /2008/oct/21/forests-conservation; Woodland Trust, *The Current State of Ancient Woodland Restoration,* January 2018, https://www.woodlandtrust .org.uk/mediafile/100229275/state-of-uk-forest-report.pdf?cb=58d97f320c.

4 old oaks: Oaks are currently threatened with an alarming rise in acute oak decline. "Environmental pressures such as climate change, pollution and drought can make our oak trees more vulnerable to pests and diseases." Na- tional Trust, "A new partnership to protect oak trees from disease," https:// www.nationaltrust.org.uk/features/a-new-partnership-to-protect-oak-trees -from-disease.

4 How many more species of bird: P. Barkham, "Eight bird species are first

confirmed avian extinctions this decade," *The Guardian,* 4 September 2018, https://www.theguardian.com/environment/2018/sep/04/first-eight-bird -extinctions-of-the-21st-century-confirmed.

4   And what would this "biological annihilation": In the same week that the last male northern white rhinoceros died and two studies detailed the cata-strophic collapse of France's bird population—a third lost in fifteen years due to the disappearance of the insects they feed on—a UN-backed report argued that biodiversity loss should be considered as alarming as climate change. The study of studies from over 100 countries, by over 550 experts, found that the destruction of nature will affect people alive today. Biodiversity loss is often talked about in terms of how our children's children will cope, but the report pulled the risks back from the abstract future. For example, it predicted that the fisheries in the Asia-Pacific region will decline to zero by 2048. J. Watts, "Destruction of nature as dangerous as climate change, scientists warn," *The Guardian,* 23 March 2018.

4   Around that time, I read: Robert Michael Pyle, "The Extinction of Expe-rience," http://www.morning-earth.org/DE6103/Read%20DE/Extinction %20of%20Experience.pdf.

4   "Its premise involves": Ibid.

4   In Britain, half of our: Woodland Trust, Report and Accounts, December 2017, https://www.woodlandtrust.org.uk/mediafile/100822176/wt-report-and -accounts-2017.pdf.

4   More than one in ten: RSPB, "State of nature, 2013," https://ww2.rspb.org.uk /Images/stateofnature_tcm9-345839.pdf.

5   Over just the last fifty years: D. Carrington, "Why are insects in decline, and can we do anything about it?" *The Guardian,* 10 February 2019, https://www .theguardian.com/environment/2019/feb/10/why-are-insects-in-decline-and -can-we-do-anything-about-it.

5   We live in cubicles: The estimate for an industrialized nation. See John Spengler, Akira Yamaguchi Professor of Environmental Health and Human Habitation, Harvard T. H. Chan School of Public Health, https://www.hsph .harvard.edu/john-spengler/.

5   "It describes the human costs": Richard Louv, *Last Child in the Woods* (Algonquin Books of Chapel Hill, 2008), p. 36.

5   In the same decade: http://psychoterratica.com/.

5   The environmental writer: Robin Wall Kimmerer, *Braiding Sweetgrass: Indigenous Wisdom, Scientific Knowledge and the Teachings of Plants* (Milk-weed Editions, 2013), pp. 208–9.

5   And yet, judging by the way: R. Mason, "David Cameron at centre of 'get rid of all the green crap' storm," *The Guardian,* 21 November 2013, https:// www.theguardian.com/environment/2013/nov/21/david-cameron-green-crap -comments-storm.

6   The writer and naturalist: Richard Mabey, *Nature Cure* (Vintage, 2008), p. 37.

6   millennials, in particular: Alice Hancock, "Houseplants enjoy a growth spurt in popularity," *Financial Times,* 27 April 2018, https://www.ft.com/content /e099b9ce-43c5-11e8-803a-295c97e6fd0b.

6   Pantone's colour of the year: https://store.pantone.com/uk/en/color-of-the -year-2017/.

7   The ancient Sumerian myth: Jean Delumeau, *History of Paradise: The Garden of Eden in Myth and Tradition,* translated by Matthew O'Connell (University of Illinois Press, 2000), p. 5.

7   Early Sanskrit literature: Quoted by Henry David Thoreau in *Walden* (1854; Penguin Classics, 2016), p. 82; translated, according to the note there, from the *Harivamsa.*

7   In the pre-industrialized West: Francesco Petrarca, "The Ascent of Mont Ventoux," in *The Renaissance Philosophy of Man,* ed. E. Cassirer et al. (University of Chicago Press, 1948), pp. 36–46.

8   As the contemporary environmental historian: Roderick Nash, *Wilderness and the American Mind* (Yale University Press, 1982), p. 343.

8   The 1827 Code of Practice: Kathleen Jones, *Lunacy, Law, and Conscience, 1744–1845: The Social History of the Care of the Insane* (Routledge & Paul, 1955), p. 139; and Kathleen Jones, *Asylums and After. A Revised History of the Mental Health Services: From the Early 18th Century to the 1990s* (Athlone Press, 1993), p. 62, https://historyofmassachusetts.org/history-of-danvers-state -hospital/.

8   "I have seen, in fevers": Florence Nightingale, *Notes on Nursing: What It Is, and What It Is Not* (1859; this edition, Barnes & Noble, 2003), p. 46.

9   Others point out the global increase: K. M. Williams, E. Kraphol, E. Yonova-Doing et al., "Early life factors for myopia in the British Twins Early Development Study," *British Journal of Ophthalmology,* published online, 6 November 2018, https://bjo.bmj.com/content/early/2018/10/03/bjophthalmol-2018 -312439.info; and N. Davis, "Children urged to play outdoors to cut risk of shortsightedness," *The Guardian,* 6 November 2018, https://www.the guardian.com/science/2018/nov/06/children-urged-play-outdoors-cut-risk -shortsightedness.

9   Others worry about the "pandemic": K. D. Cashman, K. G. Dowling, Z. Škrabáková et al., "Vitamin D deficiency in Europe: pandemic?" *American Journal of Clinical Nutrition* 103(4) (2016), pp. 1033–44, https://www.ncbi.nlm.nih.gov /pmc/articles/PMC5527850/.

9   Some point to the mental health crisis: In 2014 and 2015 the female suicide rate increased to the highest rate in a decade. The incidence of self-harm among teenage girls increased by 68 percent between 2011 and 2014. Children and young teenagers are experiencing more complex mental health problems, with one in four young people (aged fifteen to twenty-four) suffering from a mental disorder. Eating disorders are more common than ever before. Since 1993, the prevalence of common mental disorders (depression, generalized anxiety disorders) has risen by one-fifth.

9    "deaths of despair": A. Case and A. Deaton, "Mortality and morbidity in the 21st century," *Brookings Papers on Economic Activity,* Spring 2017, https://www.brookings.edu/wp-content/uploads/2017/08/casetextsp17bpea.pdf.

9    In the United States, the suicide rate: R. Prasad, "Why US suicide rate is on the rise," *BBC News,* 11 June 2018, https://www.bbc.co.uk/news/world-us-canada-44416727.

9    experts point to the growing: A. Mohdin, "Suicide rate rises among young people in England and Wales," *The Guardian,* 4 September 2018, https://www.theguardian.com/society/2018/sep/04/suicide-rate-rises-among-young-people-in-england-and-wales.

14   Looking beneath the hood of: Charles E. Beveridge, "Frederick Law Olmsted's theory on landscape design," *Nineteenth Century* 3 (Summer 1977), pp. 38–43, quoted in D. Schuyler and G. Kaliss (eds.), *The Papers of Frederick Law Olmsted,* vol. 9: *The Last Great Projects, 1890–1895* (Johns Hopkins University Press, 2015), p. 528, n. 8.

## 1 · OLD FRIENDS

18   In 2004, Mary O'Brien: M. E. R. O'Brien, H. Anderson, E. Kaukel, K. O'Byrne, M. Pawlicki, J. von Pawel and M. Reck, on behalf of SR-ON-12 Study Group, "SRL172 (killed *Mycobacterium vaccae*) in addition to standard chemotherapy improves quality of life without affecting survival, in patients with advanced non-small-cell lung cancer: phase III results," *Annals of Oncology* 15(6) (June 2014), pp. 906–14, https://www.ncbi.nlm.nih.gov/pubmed/15151947.

18   She wanted to see if: C. Abou-Zeid et al., "Induction of a type 1 immune response to a recombinant antigen from *Mycobacterium tuberculosis* expressed in *Mycobacterium vaccae,*" *Infection and Immunity* 65(5) (May 1997), pp. 1856–62, https://iai.asm.org/content/65/5/1856.short.

18   Separately, a neuroscientist: C. A. Lowry et al., "Identification of an immune-responsive mesolimbocortical serotonergic system: potential role in regulation of emotional behavior," *Neuroscience* 146(2) (2007), pp. 756–72, doi:10:1016/j.neuroscience.2007:01.067.

19   And it did: Stefan O. Reber et al., "Immunization with a heat-killed preparation of the environmental bacterium *Mycobacterium vaccae* promotes stress resilience in mice," *Proceedings of the National Academy of Sciences,* 113(22) (May 2016), pp. E3130–E3139, doi:10:1073/pnas.1600324113, http://www.pnas.org/content/113/22/E3130/.

19   You almost certainly have microscopic: There are two species of mite that live on your face: *Demodex folliculorum* and *D. brevis.* L. Jones, "These microscopic mites live on your face," *BBC Earth,* 8 May 2015.

19   That figure has been downgraded: R. Sender, S. Fuchs and R. Milo, "Revised estimates for the number of human and bacteria cells in the body," *PLoS Biology* 14(8) (August 2016), e1002533, doi:10:1371/journal.pbio.1002533, https://www.ncbi.nlm.nih.gov/pmc/articles/PMC4991899/.

19   These organisms aren't simply parasitic: W. A. de Steenhuijsen Piters, E. A. Sanders and D. Bogaert, "The role of the local microbial ecosystem in respiratory health and disease," *Philosophical Transactions of the Royal Society of London, Series B, Biological Sciences* 370(1675) (2015), pii: 20140294, doi:10.1098/rstb.2014:0294, https://www.ncbi.nlm.nih.gov/pmc/articles/PMC 4528492/.

19   Studies suggest fifty different: L. Macovei, J. McCafferty, T. Chen, F. Teles, H. Hasturk, B. J. Paster and A. Campos-Neto, "The hidden 'mycobacteriome' of the human healthy oral cavity and upper respiratory tract," *Journal of Oral Microbiology* 7(1) (February 2015), https://www.ncbi.nlm.nih.gov/pubmed /25683180.

20   But the human epidermis: Paul Shepard, "Ecology and man—a viewpoint" (1969), in Lorne J. Forstner and John H. Todd (eds.), *The Everlasting Universe: Readings on the Ecological Revolution* (D. C. Heath and Company, 1971), https://paulhoweshepard.wordpress.com/quoteexcerpts/.

20   This is the simmering, low-level: M. Anft, "Understanding inflammation," *Johns Hopkins Health Review* 3(1) (Spring/Summer 2016), https:// www.johnshopkinshealthreview.com/issues/spring-summer-2016/articles /understanding-inflammation.

21   As the neuropsychiatrist: *BBC News,* "Can mental illness start in your immune system?" 5 June 2018, https://www.bbc.co.uk/programmes/articles /5XdWrYsyCWk3YS4Vhl8V4ob/can-mental-illness-start-in-your-immune -system.

21   A study of fifteen thousand children: Edward Bullmore, *The Inflamed Mind* (Short Books Ltd., 2018), p. 12.

21   People with depression, anxiety: Ibid., p. 116.

21   European people have higher levels: Ibid., p. 149.

21   Early findings suggest: N. Kappelmann, G. Lewis, R. Dantzer, P. B. Jones and G. M. Khandaker, "Antidepressant activity of anti-cytokine treatment: a systematic review and meta-analysis of clinical trials of chronic inflammatory conditions," *Molecular Psychiatry* 23 (2018), pp. 335–43, https://www.nature .com/articles/mp2016167.

21   People with a dysregulated: S. M. Gibney and H. A. Drexhage, "Evidence for a dysregulated immune system in the etiology of psychiatric disorders," *Journal of Neuroimmune Pharmacology* 8(4) (September 2013), pp. 900–920, https://www.ncbi.nlm.nih.gov/pubmed/23645137.

21   Studies show that just two hours: S. G. Im, H. Choi, Y.-H. Jeon, M.-K. Song, W. Kim and J.-M. Woo, "Comparison of effect of two-hour exposure to forest and urban environments on cytokine, anti-oxidant, and stress levels in young adults," *International Journal of Environmental Research and Public Health* 13(7) (June 2016), p. 625, https://www.ncbi.nlm.nih.gov/pmc/articles /PMC4962166/.

22   Rook has spent much of his: G. A. Rook, "Regulation of the immune system by biodiversity from the natural environment: an ecosystem service essential

to health," *Proceedings of the National Academy of Sciences* 110(46) (November 2013), http://www.grahamrook.net/resources/Rook_PNAS-2013-1313731110.pdf.

22 "Not only are they less likely": It is worth noting that farmers in the UK are at a higher risk of suicide than other professional groups. Factors include rural decline, access to firearms, financial difficulties and a functional attitude to death and isolation. N. Booth, M. Briscoe and R. Powell, "Suicide in the farming community: methods used and contact with health services," *Occupational and Environmental Medicine* 57(9) (2000), pp. 642–44, https://oem.bmj.com/content/57/9/642.

23 They live across the United States: Young Center for Anabaptist and Pietist Studies, Elizabethtown College, "Amish population change, 2010–2015," https://groups.etown.edu/amishstudies/files/2016/06/Population_Change_2010-2015.pdf.

23 A study of the innate immunity: M. M. Stein, C. L. Hrusch, J. Gozdz et al., "Innate immunity and asthma risk in Amish and Hutterite farm children," *New England Journal of Medicine* 375(5) (2016), pp. 411–21, doi:10:1056/NEJMoa1508749, https://www.ncbi.nlm.nih.gov/pmc/articles/PMC5137793/.

24 Also, there are permanently: Rook, "Regulation of the immune system by biodiversity from the natural environment."

24 By 2050, 68 percent: United Nations, *World Urbanization Prospects: The 2018 Revision* (ST/ESA/SER.A/420), UN Department of Economic and Social Affairs, Population Division, 2019, https://population.un.org/wup/Publications/Files/WUP2018.Report.pdf.

24 In 2018, a study group: Till S. Böbel, Sascha B. Hackl, Dominik Langgartner, Marc N. Jarczok, Nicolas Rohleder, Graham A. Rook, Christopher A. Lowry, Harald Gündel, Christiane Waller and Stefan O. Reber, "Less immune activation following social stress in rural vs. urban participants raised with regular or no animal contact, respectively," *Proceedings of the National Academy of Sciences* 115(20) (May 2018), https://www.pnas.org/content/115/20/5259.

25 In 2007, chemists at Brown: Brown University website, "Brown chemists explain the origin of soil-scented geosmin," 16 September 2007, https://news.brown.edu/articles/2007/09/origin-soil-scented-geosmin.

26 A paper by a team at the School: M. Kim, K. Sowndhararajan, T. Kim, J. E. Kim, J. E. Yang and S. Kim, "Gender differences in electroencephalographic activity in response to the earthy odorants geosmin and 2-methylisoborneol," *Applied Sciences* 7(9) (2017), p. 876, https://www.mdpi.com/2076-3417/7/9/876/htm.

## 2 · BIOPHILIA

30 The biophilia concept: Wilson put forward the hypothesis in 1984, and then expanded it in 1993 with *The Biophilia Hypothesis,* a collection of essays by academics across various disciplines, edited by Wilson and the late Stephen R. Kellert, Professor of Social Ecology at Yale.

30 He argues that the biophilic: Edward O. Wilson, *Biophilia* (Harvard University Press, 1984), p. 85.

30 "Snakes mattered": Ibid., p. 101.

31 "A certain genotype makes": Edward O. Wilson, "Biophilia and the conservation ethic," in Stephen R. Kellert and Edward O. Wilson (eds.), *The Biophilia Hypothesis* (Island Press, 1993), p. 33.

31 In other words, "we learn": Wilson, *Biophilia,* p. 106.

31 The landscape architects and gardeners: Ibid., p. 111.

32 "These features are all": Judith H. Heerwagen and Gordon H. Orians, "Humans, habitats, and aesthetics," in Kellert and Wilson (eds.), *The Biophilia Hypothesis,* p. 155.

32 At the International Conference: Muhamad Solehin Fitry Rosley, Syumi Rafida Abdul Rahman and Hasanuddin Lamit, "Biophilia theory revisited: experts and non-experts perception on aesthetic quality of ecological landscape," *Procedia—Social and Behavioral Sciences* 153 (16 October 2014), pp. 349–62, https://www.sciencedirect.com/science/article/pii/S1877042814055104.

34 "The results from our preliminary": Heerwagen and Orians, "Humans, habitats, and aesthetics," p. 160. The group ranged in ages from eighteen to sixty outside the University of Washington Bookstore (72) and a restaurant on the university's campus (30).

34 In tribute to the recently deceased: Guy Lochhead, "Prospect-refuge theory," *Ernest,* 3 August 2015, http://www.ernestjournal.co.uk/blog/2015/7/28/prospect-refuge-theory.

35 People "prefer entities that are complicated": Wilson, *Biophilia,* p. 116.

35 A study of newborn babies: Francesca Simion, Lucia Regolin and Hermann Bulf, "A predisposition for biological motion in the newborn baby," *Proceedings of the National Academy of Sciences* 105(2) (January 2008), pp. 809–13, http://www.pnas.org/content/105/2/809.

35 The fact that we are attracted: Wilson, *Biophilia,* p. 113.

35 A landscape isn't just nice: Ibid.

36 Roger Ulrich, Professor of Architecture: R. S. Ulrich, "View through a window may influence recovery from surgery," *Science* 224(4647) (April 1984), pp. 420–21, https://www.ncbi.nlm.nih.gov/pubmed/6143402.

37 Of course, it's not as simple: J. J. Alvarsson, S. Wiens and M. E. Nilsson, "Stress recovery during exposure to nature sound and environmental noise," *International Journal of Environmental Research and Public Health* 7(3) (2010), pp. 1036–46, https://www.ncbi.nlm.nih.gov/pmc/articles/PMC2872309/.

37 From October 2018, doctors: S. Carrell, "Scottish GPs to begin prescribing rambling and birdwatching," *The Guardian,* 5 October 2018, https://www.theguardian.com/uk-news/2018/oct/05/scottish-gps-nhs-begin-prescribing-rambling-birdwatching.

38 In the United States, between April: Park Rx America–Leaderboard, https://parkrxamerica.org/leaderboard/.

38   In an essay for *The Biophilia:* Madhav Gadgil, "Of life and artifacts," in Kellert and Wilson (eds), *The Biophilia Hypothesis,* p. 366.

39   In 2009, an evaluation: B. Grinde and G. G. Patil, "Biophilia: does visual contact with nature impact on health and well-being?" *International Journal of Environmental Research and Public Health* 6(9) (2009), pp. 2332–43, https://www.ncbi.nlm.nih.gov/pmc/articles/PMC2760412/#b54-ijerph-06 -02332.

40   Indeed, look at the way: "So Charles was right—you should talk to plants, scientists discover," *Evening Standard,* 29 August 2007, https://www.standard .co.uk/news/so-charles-was-right-you-should-talk-to-plants-scientists-discover -6609430.html.

40   In a later interview: "I talk to plants, but that doesn't mean I'm potty, says Charles," *Sydney Morning Herald,* 21 September 2010, https://www.smh.com .au/entertainment/celebrity/i-talk-to-plants-but-that-doesnt-mean-im-potty -says-charles-20100920-15jor.html.

40   Nowadays, it is not so clear: Wilson, *Biophilia,* p. 116.

40   It was the "extinction": Robert Michael Pyle, "The Extinction of Experience"; see Introduction, note at "Around that time, I read."

40   Wilson feels strongly about this: Wilson, *Biophilia,* p. 118.

### 3 · MUD-LUSCIOUS AND PUDDLE-WONDERFUL

45   Three-quarters of children: D. Carrington, "Three-quarters of UK children spend less time outdoors than prison inmates—survey," *The Guardian,* 25 March 2016, https://www.theguardian.com/environment/2016/mar/25/three -quarters-of-uk-children-spend-less-time-outdoors-than-prison-inmates-survey.

45   We've moved far away from: Jean-Jacques Rousseau, *Émile* (Dover Publications, 2013), p. 30.

45   Fewer than one in ten: Stephen Moss, *Natural Childhood,* National Trust report, 2012, https://www.nationaltrust.org.uk/documents/read-our-natural -childhood-report.pdf.

45   Break-time or recess: Peter Blatchford and Clare Sumpner, "What do we know about breaktime? Results from a national survey of breaktime and lunchtime in primary and secondary schools," *British Educational Research Journal* 24(1) (1998), pp. 79–94, https://www.tandfonline.com/doi/abs/10 :1080/0141192980240106.

45   and the United States: https://usplaycoalition.org/wpcontent/uploads/2015/ 08/13:11.5_Recess_final_online.pdf.

45   In 2013, the RSPB published: RSPB, *Connecting with Nature,* University of Essex, 2013, http://ww2.rspb.org.uk/Images/connecting-with-nature_tcm9-354 603.pdf.

45   If a child is introduced: "Children who don't connect with nature before the age of 12 are less likely as adults to connect with nature. They therefore lose

out on the resilience nature provides when you're really stressed." Dr. William Bird, Outdoor Nation Interview, quoted in Moss, *Natural Childhood*.

45 Secondly, without exposure: Dr. William Bird, *Natural Thinking,* Report for the RSPB, June 2007, http://ww2.rspb.org.uk/Images/naturalthinking _tcm9-161856.pdf.

45 Although children spend much: R. Jenkins, "Children spend twice as long looking at screens as they do playing outside, a study has found," *The Independent,* 26 October 2018, https://www.independent.co.uk/life-style/children -screens-play-outside-computer-phone-time-healthy-games-a8603411.html.

46 Newcastle council has cut: D. Kirby, "Newcastle parks could be used to bury bodies under proposal to raise funds for charitable trust," *iNews* (newspaper), 21 November 2017, https://inews.co.uk/news/uk/newcastle-parks-used-bury -bodies-proposal-raise-funds-charitable-trust/.

46 "Let us remember that it": Quoted in Robin G. Schulze, *The Degenerate Muse: American Nature, Modernist Poetry, and the Problem of Cultural Hygiene* (Oxford University Press, 2013), p. 55.

47 At the expense of outdoor play: Alison Flood, "Oxford Junior Dictionary's replacement of 'natural' words with 21st-century terms sparks outcry," *The Guardian,* 13 January 2015, https://www.theguardian.com/books/2015/jan /13/oxford-junior-dictionary-replacement-natural-words.

47 "It isn't the job": Martin Robbins, "Why Oxford Dictionaries are right to purge nature from the dictionary," *The Guardian,* 3 March 2015, https:// www.theguardian.com/science/the-lay-scientist/2015/mar/03/why-the-oed -are-right-to-purge-nature-from-the-dictionary.

47 As the writer Jay Griffiths: Jay Griffiths, *Kith: The Riddle of the Childscape* (Penguin Books, 2014), p. 253.

47 As Robin Wall Kimmerer puts it: Robin Wall Kimmerer, *Braiding Sweetgrass: Indigenous Wisdom, Scientific Knowledge and the Teachings of Plants* (Milkweed Editions, 2013), p. 57.

48 A 1998 study in Chicago: A. F. Taylor, A. Wiley, F. E. Kuo and W. C. Sullivan, "Growing up in the inner city: green spaces as places to grow," *Environment and Behavior* 30(1) (1998), pp. 3–27, https://is.muni.cz/el/1423/ jaro2014/HEN597/um/47510652/Faber_Taylor__A.__Wiley__AKuo__ F._E.___Sullivan__W._C.__1998_.pdf.

48 "It is necessary to be outside": *Nature is the Perfect Playground,* Friluftsfrämjandet organization, 2015, http://igniteup.co.uk/wp-content/uploads/2015 /04/Skogsmulle-leaflet.pdf.

48 "especially between the ages": J. Bogousslavsky, "In memoriam: David H. Ingvar," *Cerebrovascular Diseases* 11(1) (January 2001), https://www.karger .com/Article/PDF/47614.

49 And yet, amazingly: "Advice on standards for school premises: for local authorities, proprietors, school leaders, school staff and governing bodies," Department for Education, March 2015, https://assets.publishing.service

.gov.uk/government/uploads/system/uploads/attachment_data/file/410294
/Advice_on_standards_for_school_premises.pdf.

49    "Much of society—including": Personal interview with Richard Louv, by
email, 6 July 2017.

49    "A playground with a lot of": Fredrika Mårtensson, "Children and nature,"
in Matilda van den Bosch and William Bird (eds.), *Oxford Textbook of Nature
and Public Health* (Oxford University Press, 2018), Chapter 6:1, Box 6:1.1,
p. 168.

49    In 2007, UNICEF: Ipsos MORI and Dr. Agnes Nairn, *Child Well-being in
the UK, Spain and Sweden: The Role of Inequality and Materialism,* report for
UNICEF, 2007, https://downloads.unicef.org.uk/wp-content/uploads/2011
/09/UNICEFIpsosMori_childwellbeing_reportsummary.pdf.

49    Another study tracked people: "Scouts and guides provide 'mental health
boost for life,'" *BBC News,* 10 November 2016, https://www.bbc.co.uk/news
/uk-scotland-37923133.

49    Access to nature was found: N. M. Wells and G. W. Evans, "Nearby nature:
a buffer of life stress among rural children," *Environment and Behavior* 35(3)
(May 2003), pp. 311–30, https://pdfs.semanticscholar.org/f1b3/b8b51f9b1129
5debee2b9b4956e24422e6f9.pdf.

50    What is so striking and worrying: A. Oudin, L. Bråbäck, D. O. Åström et
al., "Association between neighbourhood air pollution concentrations and
dispensed medication for psychiatric disorders in a large longitudinal cohort
of Swedish children and adolescents," *BMJ Open* 6(6) (2016), https://bmjopen
.bmj.com/content/6/6/e010004.full.

50    The EU and WHO: Greenpeace, "New satellite data reveals world's largest
NO2 air pollution emission hotspots," *Greenpeace Media Briefing,* October
2018, https://storage.googleapis.com/planet4-international-stateless/2018/10
/07426a79-no2-air-pollution-analysis-greenpeace.pdf.

50    It makes decisions to: M. Kinver, "Growth of city trees can cut air pollution, says report," *BBC News,* 31 October 2016, https://www.bbc.co.uk/news
/science-environment-37813709.

50    Sheffield, Newcastle and Edinburgh: A. Martin, "Great tree massacre: councils have cut down 110,000 trees in a desperate bid to save money," *Daily Mail,*
3 June 2018, https://www.dailymail.co.uk/news/article-5801627/Great-tree
-massacre-Councils-cut-110-000-trees-desperate-bid-save-money.html.

50    In 2018, a team: Carla P. Bezold et al., "The association between natural
environments and depressive symptoms in adolescents living in the United
States," *Journal of Adolescent Health* 62(4) (April 2018), pp. 488–95, https://
www.hsph.harvard.edu/news/hsph-in-the-news/greenery-depression-teens/.

51    We now know that air pollution: D. Carrington, "Air pollution particles
found in foetal side of placentas—study," *The Guardian,* 17 September 2019.

51    In 2019, a team at King's: J. B. Newbury, L. Arseneault, S. Beevers et al.,
"Association of air pollution exposure with psychotic experiences during ado-

lescence," *JAMA Psychiatry* 76(6) (2019), pp. 614–23, https://jamanetwork
.com/journals/jamapsychiatry/fullarticle/2729441.

51   Thousands of children in Britain: S. Marsh, "Calls for action over UK's
'intolerable' child mental health crisis," *The Guardian,* 31 August 2018,
https://www.theguardian.com/society/2018/aug/31/calls-for-action-over-uks
-intolerable-child-mental-health-crisis.

51   As the epidemiologist: Personal interview with Howard Frumkin, 21 May
2018.

53   It is simple and old-fashioned: P. Cocozza, "Are iPads and tablets bad for
young children?," *The Guardian,* 8 January 2014, https://www.theguardian
.com/society/2014/jan/08/are-tablet-computers-bad-young-children.

55   As concern grows in Britain: In September 2012, the World Congress of the
International Union for the Conservation of Nature (IUCN) recognized a
"widely shared concern about the increasing disconnection of people and especially children from nature, and the adverse consequences for both healthy
child development ('nature deficit disorder') as well as responsible stewardship for nature and the environment in the future," https://www.ohchr.org
/Documents/HRBodies/CRC/Discussions/2016/AnneliesHenstra_en.pdf.

55   Among parents, there is an: http://www.natureschools.org.uk/.

55   Its priority, it said: Department for Education, response to petition: "Develop
a GCSE in Natural History," July 2017, https://petition.parliament.uk
/archived/petitions/176749.

55   Over a four-year period: Alan Williams, "England's largest outdoor learning
project reveals children more motivated to learn when outside," press release,
July 2016, about Natural Connections Demonstration project, funded by
Natural England, the Department for the Environment, Food and Rural
Affairs and Historic England and delivered by the University of Plymouth,
https://www.plymouth.ac.uk/news/englands-largest-outdoor-learning-project
-reveals-children-more-motivated-to-learn-when-outside.

56   Studies of the long-term impact: S. Blackwell, *Impacts of Long Term Forest
School Programmes on Children's Resilience, Confidence and Wellbeing,* June
2015, https://getchildrenoutdoors.files.wordpress.com/2015/06/impacts-of-long
-term-forest-schools-programmes-on-childrens-resilience-confidence-and
-wellbeing.pdf.

56   "It stopped low-level disruption": Personal interview with Fred Banks, 18 July
2017.

56   Government and schools should: Michael Jonathan Klassen, "Connectedness to nature: comparing rural and urban youths' relationships with
nature," Royal Roads University, Victoria, Canada, MA thesis, January 2010,
https://viurrspace.ca/bitstream/handle/10170/150/Klassen%2C%20Mike.pdf
?sequence=1&isAllowed=y.

57   In the United Kingdom, people from: J. E. Booth, K. J. Gaston and P. R. Arms-
worth, "Who benefits from recreational use of protected areas?," *Ecology and*

*Society* 15(3) (2010), https://www.ecologyandsociety.org/vol15/iss3/art19/; The Countryside Agency, *Capturing Richness: Countryside Visits by Black and Ethnic Minority Communities,* September 2003, https://webarchive .nationalarchives.gov.uk/20120302162614/; http://www.naturalengland.org .uk/Images/capturingrichnessfinaltcm2-10023_tcm6-4017.pdf.

58  In June 2018 a study: Rosemary R. C. McEachan et al., "Availability, use of, and satisfaction with green space, and children's mental wellbeing at age 4 years in a multicultural, deprived, urban area: results from the 'Born in Bradford' cohort study," *The Lancet Planetary Health* 2(6) (June 2018), pp. e244–e254.

58  It found that greater satisfaction: There was no association for white British children, which the team said warranted further study.

58  "This is an important finding": Jenny Roe, "Ethnicity and children's mental health: making the case for access to urban parks," *The Lancet Planetary Health,* Comment 2(6) (June 2018), pp. e234–e235.

## 4 · PHYSIOLOGICAL RESONANCE

65  In the 1990s, a psychologist: Summer Allen, "The science of awe," Greater Good Science Center at UC Berkeley for the John Templeton Foundation, September 2018, https://ggsc.berkeley.edu/images/uploads/GGSC-JTF_White _Paper-Awe_FINAL.pdf.

65  "Awe binds us to social": D. Keltner, "Why do we feel awe?," *Greater Good Magazine,* 10 May 2016, https://greatergood.berkeley.edu/article/item/why _do_we_feel_awe.

65  This idea was borne out: Paul K. Piff, Pia Dietze, Matthew Feinberg, Daniel M. Stancato, Dacher Keltner, "Awe, the small self, and prosocial behavior," *Journal of Personality and Social Psychology* 108(6) (2015), pp. 883–99, http:// psycnet.apa.org/buy/2015-21454-002.

65  Professor Jennifer Stellar: J. E. Stellar, N. John-Henderson, C. L. Anderson, A. M. Gordon, G. D. McNeil and D. Keltner, "Positive affect and markers of inflammation: discrete positive emotions predict lower levels of inflammatory cytokines," *Emotion* 15(2) (April 2015), pp. 129–33, https://www.ncbi.nlm.nih .gov/pubmed/25603133.

66  After white-water rafting: C. L. Anderson, M. Monroy and D. Keltner, "Awe in nature heals: evidence from military veterans, at-risk youth, and college students," *Emotion* 18(8) (December 2018), pp. 1195–202, https://www.ncbi.nlm .nih.gov/pubmed/29927260.

68  In 1980, Rachel and Stephen: S. Kaplan, "The restorative benefits of nature: toward an integrative framework," *Journal of Environmental Psychology* 15(3) (September 1995), pp. 169–82, http://willsull.net/resources/KaplanS1995.pdf.

68  "If you can find an environment": Rebecca A. Clay, "Green is good for you," *APA Monitor* 32(4) (April 2001), p. 40 (print), http://www.apa.org/monitor /apr01/greengood.aspx.

69  The scientists concluded that: Kate E. Lee, Kathryn J. H. Williams, Leisa

D. Sargent, Nicholas S. G. Williams and Katherine A. Johnson, "40-second green roof views sustain attention: the role of micro-breaks in attention restoration," *Journal of Environmental Psychology* 42 (June 2015), pp. 182–89, https://www.sciencedirect.com/science/article/pii/S0272494415000328.

69    The study author Dr. Kate Lee: "It's really important to have micro-breaks. It's something that a lot of us do naturally when we're stressed or mentally fatigued," study author Dr. Kate Lee said. "There's a reason you look out the window and seek nature, it can help you concentrate on your work and to maintain performance across the workday." Lee, Williams et al., "40-second green roof views sustain attention." See also University of Melbourne, "Glancing at greenery on a city rooftop can markedly boost concentration levels," *ScienceDaily,* 26 May 2015, https://www.sciencedirect.com/science/article/abs/pii/S0272494415000328.

69    In 2009, a study from researchers: Andrea Faber Taylor and Frances E. Kuo, "Children with attention deficits concentrate better after walk in the park," *Journal of Attention Disorders* 12(5) (March 2009), pp. 402–9, http://journals.sagepub.com/doi/abs/10.1177/1087054708323000.

70    Electroencephalography (EEG): EEGs record how excited the neurons— specialized brain cells—are and the extent to which they fire together. There are four types of brain waves an EEG can pick up. Alpha waves suggest a state of relaxation and creativity. Beta waves are found when a person is carrying out a focused task. Delta waves are associated with sleep. Gamma waves are associated with consciousness, universal love and spirituality, depending on who you talk to.

70    Cerebral activity in: B.-J. Park, Y. Tsunetsugu, T. Kasetani, H. Hirano, T. Kagawa, M. Sato and Y. Miyazaki, "Physiological effects of Shinrin-yoku (taking in the atmosphere of the forest)—using salivary cortisol and cerebral activity as indicators," *Journal of Physiological Anthropology* 26(2) (March 2007), pp. 123–28, http://www.ncbi.nlm.nih.gov/pubmed/17435354.

70    Your body may produce: Salivary cortisol is used as a biomarker—a biological marker that is evaluated and measured to indicate a biological or pathological process, or a response to a treatment, particularly reaction to stress.

70    There may also be reduced: G. N. Bratman, G. C. Daily, B. J. Levy and J. J. Gross, "The benefits of nature experience: improved affect and cognition," *Landscape and Urban Planning* 138 (2015), pp. 41–50, https://www.sciencedirect.com/science/article/pii/S0169204615000286.

70    In 1947, Donald O. Hebb: Matilda van den Bosch and William Bird (eds.), *Oxford Textbook of Nature and Public Health* (Oxford University Press, 2018), Chapter 2:4, p. 71.

70    The group, led by the American: M. R. Rosenzweig, D. Krech, E. L. Bennett and M. C. Diamond, "Effects of environmental complexity and training on brain chemistry and anatomy: a replication and extension," *Journal of Comparative Physiological Psychology* 55 (1962), pp. 429–37.

71    Scientists hypothesized that: M. R. Rosenzweig, "Modification of brain cir-

cuits through experience," in F. Bermúdez-Rattoni (ed.), *Neural Plasticity and Memory: From Genes to Brain Imaging* (CRC Press/Taylor & Francis, 2007), Chapter 4.

71 Another concept, the "inoculation": E. J. Crofton, Y. Zhang and T. A. Green, "Inoculation stress hypothesis of environmental enrichment," *Neuroscience & Biobehavioral Reviews* 49 (2015), pp. 19–31, https://www.ncbi.nlm.nih.gov /pmc/articles/PMC4305384.

71 These mild stressors trigger: "How nature can affect health—theories and mechanisms," section 2 in Van den Bosch and Bird (eds.), *Oxford Textbook of Nature and Public Health,* p. 74.

72 According to a 2014 report: Emily Hewlett and Valerie Moran, *Making Mental Health Count: The Social and Economic Costs of Neglecting Mental Health Care* (OECD, 2014), https://read.oecd-ilibrary.org/social-issues -migration-health/making-mental-health-count_9789264208445-en#.WoG EOZPFLLY.

72 (such as accessible health): D. Vlahov, S. Galea and N. Freudenberg, "The urban health 'advantage,'" *Journal of Urban Health: Bulletin of the New York Academy of Medicine* 82(1) (2005), https://www.ncbi.nlm.nih.gov/pmc /articles/PMC3456628/pdf/11524_2006_Article_339.pdf.

72 Living in greener neighbourhoods: Carmen de Keijzer et al., "Residential sur-rounding greenness and cognitive decline: a 10-year follow-up of the White-hall II cohort," *Environmental Health Perspectives* 126(7) (July 2018), https:// ehp.niehs.nih.gov/doi/10.1289/EHP2875.

72 A neuroscientific approach: F. Lederbogen, P. Kirsch, L. Haddad, F. Streit, H. Tost, P. Schuch, S. Wüst, J. C. Pruessner, M. Rietschel, M. Deuschle and A. Meyer-Lindenberg, "City living and urban upbringing affect neural social stress processing in humans," *Nature* 474(7352) (23 June 2011), pp. 498–501, http://sa.indiaenvironmentportal.org.in/files/City%20living.pdf.

72 The study was designed to: K. Dedovic, R. Renwick, N. K. Mahani, V. Engert, S. J. Lupien and J. C. Pruessner, "The Montreal Imaging Stress Task: using functional imaging to investigate the effects of perceiving and processing psychosocial stress in the human brain," *Journal of Psychiatry & Neuroscience* 30(5) (2005), pp. 319–25, https://www.ncbi.nlm.nih.gov/pmc /articles/PMC1197276/.

72 "We know what the amygdala": Lederbogen, Kirsch et al., "City living and urban upbringing affect neural social stress processing in humans," https:// www.nature.com/articles/nature10190 and http://sa.indiaenvironmentportal .org.in/files/City%20living.pdf.

73 "There's prior evidence that": A. Jha, "City living affects your brain, research-ers find," *The Guardian,* 22 June 2011, https://www.theguardian.com/science /2011/jun/22/city-living-afffects-brain.

73 A group in Edinburgh used: S. Tilley, C. Neale, A. Patuano and S. Cin-derby, "Older people's experiences of mobility and mood in an urban envi-ronment: a mixed methods approach using electroencephalography (EEG)

and interviews," *International Journal of Environmental Research and Public Health* 14(2) (February 2017), p. e151, https://www.ncbi.nlm.nih.gov/pubmed /28165409.

74   In the 2000s, Richard Taylor: C. M. Hagerhall, T. Laike, R. P. Taylor, M. Küller, R. Küller and T. P. Martin, "Investigations of human EEG response to viewing fractal patterns," *Perception* 37(10) (2008), pp. 1488–94, http:// journals.sagepub.com/doi/10:1068/p5918.

74   different types of fractals: R. P. Taylor, B. Spehar, J. A. Wise, C. W. G. Clifford, B. R. Newell, C. M. Hagerhall, T. Purcell and T. P. Martin, "Perceptual and physiological responses to the visual complexity of fractal patterns," *Nonlinear Dynamics, Psychology, and Life Sciences* 9(1) (January 2005), pp. 89–114, https://cpbuse1.wpmucdn.com/blogs.uoregon.edu/dist/e/12535/files/2015/12 /ResponseNon-linear-28e9hbu.pdf.

76   "No self-respecting Town Council": Brent Elliott, *Occasional Papers from the RHS Lindley Library* 12 (September 2014): *Horticulture and the First World War,* https://www.rhs.org.uk/about-the-rhs/pdfs/publications/lindley -library-occasional-papers/volume-12-september-2014.pdf.

77   A cross-discipline research team: Sierra Club, Military Outdoors, https:// content.sierraclub.org/outings/military.

77   a clinical trial in 2020: S. O'Brien, "Can hiking help heal veterans with PTSD? Researchers seek to find out," *REI Co-Op Journal,* 22 October 2018, https://www.rei.com/blog/hike/can-hiking-help-heal-veterans-with-ptsd -researchers-seek-to-find-out.

77   Veterans with PTSD: Carly M. Rogers, Trudy Mallinson and Dominique Peppers, "High-intensity sports for posttraumatic stress disorder and depression: feasibility study of ocean therapy with veterans of Operation Enduring Freedom and Operation Iraqi Freedom," *American Journal of Occupational Therapy* 68(4) (July/August 2014), pp. 395–404, https://ajot.aota.org/Article .aspx?articleid=1884509.

77   In San Diego, the navy is spending: T. Perry, "Riding the waves to better health: Navy studies the therapeutic value of surfing," *Washington Post,* 10 March 2018, https://www.washingtonpost.com/national/health-science /riding-the-waves-to-better-health-navy-studies-the-therapeutic-value-of -surfing/2018/03/09/254df9e2-06ca-11e8-94e8-e8b8600ade23_story.html ?utm_term=.351bd6af3756.

77   After exposure to nature: M. M. H. E. van den Berg, J. Maas, R. Muller et al., "Autonomic nervous system responses to viewing green and built settings: differentiating between sympathetic and parasympathetic activity," *International Journal of Environmental Research and Public Health* 12(12) (December 2015), pp. 15860–74, https://www.ncbi.nlm.nih.gov/pmc/articles /PMC4690962/.

77   This has important consequences: Ibid.

77   High resting levels of parasympathetic: Ibid.

78   A meta-analysis and systematic: M. Richardson, K. McEwan, F. Maratos and

D. Sheffield, "Joy and calm: how an evolutionary functional model of affect regulation informs positive emotions in nature," *Evolutionary Psychological Science* 2(4) (December 2016), pp. 308–20.

78 They were more likely: Other studies have found similar results; particularly influential was a study by Roger Ulrich, published in 1991, which set out what is now known as Stress Reduction Theory (SRT). R. S. Ulrich et al., "Stress recovery during exposure to natural and urban environments," *Journal of Environmental Psychology* 11(3) (September 1991), pp. 201–30, https://psych .utah.edu/_documents/psych4130/Ulrich%20et%20al_1991.pdf.

78 Even people under anaesthetic: Y.-C. P. Arai, S. Sakakibara, A. Ito, K. Ohshima, T. Sakakibara, T. Nishi, S. Hibino, S. Niwa and K. Kuniyoshi, "Intra-operative natural sound decreases salivary amylase activity of patients undergoing inguinal hernia repair under epidural anesthesia," *Acta Anaesthesiologica Scandinavica* 52(7) (August 2008), pp. 987–90, https://www.ncbi .nlm.nih.gov/pubmed/18477078.

78 Bernie Krause, an American: Bernie Krause, "The voice of the natural world," TED Talk, 2013, https://www.ted.com/talks/bernie_krause_the_voice_of _the_natural_world?language=en.

78 "Fully 50 percent of": Ibid.

78 Other studies investigate: Olfaction—smell—is the oldest sense. Before creatures could see, hear or touch, they evolved a rudimentary sense to detect chemicals around them. It is thought that smells enter the brain in a different, deeper, more ancient way than other senses. They go directly to the main area of processing—the olfactory bulb—without stopping off at the thalamus on the way, like vision and hearing do. So if you smell something that could be rotten or off or poisonous, you know immediately.

78 A Japanese study found that: H. Ikei, C. Song and Y. Miyazaki, "Effects of olfactory stimulation by α-pinene on autonomic nervous activity," *Journal of Wood Science* 62(6) (2016), pp. 568–72; doi:10.1007/s10086-016-1576-1.

79 Studies of night-shift workers: J. Meikle, "Night shifts raise risk of breast cancer, says Danish research," *The Guardian,* 28 May 2012, https://www .theguardian.com/society/2012/may/28/night-shift-raises-cancer-risk-study.

79 A literature review published: A. Cuomo, N. Giordano, A. Goracci and A. Fagiolini, "Depression and vitamin D deficiency: causality, assessment, and clinical practice implications," *Neuropsychiatry* 7(5) (2017), http:// www.jneuropsychiatry.org/peer-review/depression-and-vitamin-d-deficiency -causality-assessment-and-clinical-practice-implications-12051.html.

80 Wild swimmers often report: J. Landreth, "Brrr! The joys of cold water swimming," *The Telegraph,* 13 February 2017, https://www.telegraph.co.uk/health -fitness/body/brrr-joys-cold-water-swimming/.

80 Recent studies have linked: A 2004 study of winter swimmers in Finland, where a dip in cold, natural water is a popular pastime, found a decrease in tension and fatigue in swimmers, compared with the non-swimming control group, and improvements in mood and memory after four months. The swim-

mers felt more energetic and vigorous and said that the swimming helped with aches and pains. P. Huttunen, L Kokko and V. Ylijukuri, "Winter swimming improves general well-being," *International Journal of Circumpolar Health* 63(2) (2004), pp. 140–44, https://www.tandfonline.com/doi/pdf/10:3402 /ijch.v63i2:17700.

80   Crucially, negative ions: O. Pino and F. La Ragione, "There's something in the air: empirical evidence for the effects of negative air ions (NAI) on psychophysiological state and performance," *Research in Psychology and Behavioral Sciences* 1(4) (2013), pp. 48–53. "The analysis, particularly with randomised, controlled trials suggests that NAI—negative air ions—treatment for mood disorders is in general effective with effects almost equivalent to those in other antidepressant non-pharmacotherapy trials," they wrote. "Despite the growth in clinical research, there remained a substantial gap in mental health services to translate state-of-the-art treatments and incorporate them into mainstream practice." In the 2000s and 2010s, psychiatry researchers used bright light boxes and negative ion generators to study the effects on people with depression, seasonal affective disorder and bipolar conditions. Most studies concluded that, while the treatments positively show antidepressant effects, more research is needed before they are prescribed as a treatment. The evidence is enough, though, to suggest that the chemical make-up of natural spaces has a more positive effect on human mental health than that of non-natural spaces.

80   Like the smell of cedar wood: Pino and La Ragione, "There's something in the air." Positive ions are associated with negative health effects. For example, when the Santa Ana, Sharav or Sirocco blows in—the warm, dry winds— the morbidity rate increases, people get more depressed, irritable and tense. Chemists believe this is to do with the increase in positive ions. Clinical studies have linked exposure to positive ions with an increase in stress.

81   A leading expert in the field: M. Kuo, "How might contact with nature promote human health? Promising mechanisms and a possible central pathway," *Frontiers in Psychology*, 25 August 2015, https://www.frontiersin.org/articles /10:3389/fpsyg.2015:01093/full. "Enhanced immune function is known to wholly or partially subsume 11 other pathways," Kuo writes. "Each of the following is known to enhance immune function—adiponectin, reduced air pollution, awe, normalized levels of blood glucose (as compared to elevated levels), reduced obesity, physical activity, phytoncides, better sleep, social ties, relaxation and stress reduction, and reduced immediate and long-term traumatic stress due to violence. Note that these pathways may be partially or wholly subsumed by the enhanced immune functioning pathway between nature and health; if a pathway contributes to health via both the immune system and other effects, it is partially subsumed by the immune function pathway."

82   A forest therapy group: Activity in the parasympathetic (rest-and-digest) was higher, and activity in sympathetic (fight-or-flight) nervous systems was lower,

suggesting the body was relaxed. The scientists measured concentrations of cortisol, the stress hormone, in saliva; blood pressure; heart-rate variability; levels of adrenaline and noradrenaline in urine; and pulse rate, which suggested lower stress levels and a positive effect on general health, when compared with urban environments.

82    Li also found that: Q. Li, K. Morimoto, A. Nakadai, H. Inagaki, M. Katsumata, T. Shimizu et al., "Forest bathing enhances human natural killer activity and expression of anti-cancer proteins," *International Journal of Immunopathology and Pharmacology* 20(2) (April–June 2007), pp. 3–8, http://journals .sagepub.com/doi/10.1177/039463200702002S202.

82    Other psychological benefits: Kirsten Dirksen, "Science of 'forest bathing': fewer maladies, more well-being?" YouTube, https://www.youtube.com/watch ?v=9jPNll1Ccn0.

82    "If you want to decrease": Li et al., "Forest bathing enhances human natural killer activity and expression of anti-cancer proteins."

### 5 · PLANT WISDOM

83    Ecotherapy projects are currently: In New Zealand, "green prescriptions" are increasing. In the United States, the National Park Rx initiative prescribes a diverse number of programmes connecting people with local parks to improve their mental, physical and social health. At the moment, most of these projects are one-off, small-scale and focused on getting sedentary people active.

84    He argued that "ecoalienation": Howard Clinebell, "Greening pastoral care," *Journal of Pastoral Care* 48(3) (September 1994), pp. 209–14, https://journals .sagepub.com/doi/pdf/10.1177/002234099404800301.

84    But despite early positive reports: P. Währborg, I. F. Petersson and P. Grahn, "Nature-assisted rehabilitation for reactions to severe stress and/or depression in a rehabilitation garden: long-term follow-up including comparisons with a matched population-based reference cohort," *Journal of Rehabilitation Medicine: Official Journal of the UEMS European Board of Physical and Rehabilitation Medicine* 46(3) (January 2014), https://www.research gate.net/publication/259959991_Nature-assisted_rehabilitation_for_reactions _to_severe_stress_andor_depression_in_a_rehabilitation_garden_Long -Term_follow-up_including_comparisons_with_a_matched_population -based_reference_cohort.

84    "The evidence base for what": Personal interview with Dr. Rebecca Lovell, 8 August 2017.

85    "When it comes to nature contact": Personal interview with Howard Frumkin, 21 May 2018.

85    What a GP can prescribe: In Cornwall and Devon, GPs can refer patients to "Dose of Nature" groups. In Merseyside, the Natural Health Service manages a number of projects, including forest school, gardening and mindfulness outdoors. Scores of Green Gyms—where gardening and land management activi-

ties take place—are dotted across the country. In April 2017, Bristol Mental Health was one of the first groups to start offering people eight weeks of woodland sessions in its rehabilitation service. The patients had been suffering with a wide variety of mental health problems: anxiety, depression, schizophrenia, personality disorders and psychotic experiences. Once a week, a group of up to twelve people would be taken to the Forest of Avon woods for an afternoon. The therapists and well-being workers involved worked with the principles of the Five Ways to Well-being: Connect, Be Active, Take Notice, Learn and Give.

86   But since the 1980s: J. Sempik, J. Aldridge and L. Finnis (Thrive), "Social and therapeutic horticulture: the state of practice in the UK," Centre for Child and Family Research (CCFR) Family and Environment research programme, March 2004, https://dspace.lboro.ac.uk/dspace-jspui/bitstream/2134/2930/1/Evidence8.pdf.

86   "Some of these rather large": History of Thrive, https://www.thrive.org.uk/history-of-thrive.aspx.

87   The aim of the horticultural: Referrals came from social workers, GPs or families, or people could self-refer. The placements were paid for and most people used Direct Payment, which was a funded package from social services. The senior horticultural therapist at Thrive, Jan Broady, said they were surprised to be at capacity in 2017 despite the government's recent austerity measures which cut funding.

87   or illness: Thrive helps people recovering from stroke, heart disease, illness or a difficult time in their lives. Older people and young people with special needs were also within their remit.

89   They'd also chat with: For funding purposes, Thrive had to measure the positive effects of the therapy. To do this, they used a tool called Insight, which recorded behaviours, social interactions, task engagement and cognition. Analysis of the data found that the main effective period was the first two years. After that, the intervention reached a plateau but maintained its positive benefits, rather than declining.

89   Horticultural therapists are: "Gardens and gardening for people with dementia," *Thrive,* February 2011, https://www.thrive.org.uk/Files/Documents/8%20-%20Dementia%202012.pdf.

90   Tiny variations that can: V. Bourne, "Hot on the trail of those elusive gold snowdrops," *The Telegraph,* 27 January 2017, https://www.telegraph.co.uk/gardening/how-to-grow/hot-trail-elusive-gold-snowdrops/; and https://www.theguardian.com/lifeandstyle/2015/feb/19/snowdrop-flower-craze-speculative-bubble-gardening; G. Rice, "$2,500 for a bulb: is the snowdrop craze the world's next speculative bubble?" *The Guardian,* 19 February 2015, https://www.theguardian.com/lifeandstyle/2015/feb/19/snowdrop-flower-craze-speculative-bubble-gardening.

92   "The older woman was amazed at": Personal interview with John Scull, 23 August 2017.

93    I was greeted by Sarah: All names have been changed to ensure anonymity of
      staff and service users.


                              6 · EQUIGENESIS

100   Kuo and Sullivan built on: B. Cimprich, "Attentional fatigue following breast
      cancer surgery," *Research in Nursing & Health* 15(3) (June 1992), pp. 199–207,
      https://onlinelibrary.wiley.com/doi/abs/10:1002/nur.4770150306.
101   At first, they studied how: R. L. Coley, W. C. Sullivan and F. E. Kuo, "Where
      does community grow? The social context created by nature in urban pub-
      lic housing," *Environment and Behavior* 29(4) (1997), pp. 468–94, http://
      journals.sagepub.com/doi/abs/10:1177/001391659702900402.
101   "Imagine feeling irritated, impulsive": Tina Prow, "The power of trees,"
      *The Illinois Steward* 7(4) (Winter 1999), http://lhhl.illinois.edu/media/the
      poweroftrees.htm.
101   "What we have here is a naturally": Ibid.
101   To prove the link further, Kuo: F. E. Kuo and W. C. Sullivan, "Environment
      and crime in the inner city: does vegetation reduce crime?" *Environment
      and Behavior* 33(3) (2001), pp. 343–67, http://citeseerx.ist.psu.edu/viewdoc
      /download?doi=10:1.1:644:9399&rep=rep1&type=pdf.
102   "It's nice if you can do it": Personal interview with William Sullivan, 6 March
      2018.
102   In 2018, a Philadelphia-based: C. C. Branas, E. South, M. C. Kondo,
      B. C. Hohl, P. Bourgois, D. J. Wiebe and J. M. MacDonald, "Citywide cluster
      randomized trial to restore blighted vacant land and its effects on violence,
      crime, and fear," *Proceedings of the National Academy of Sciences* 115(12) (March
      2018), pp. 2946–51, https://www.mailman.columbia.edu/public-health-now
      /news/how-reduce-crime-and-gun-violence-and-stabilize-neighborhoods
      -randomized-controlled-study.
102   The results of the intervention: Ibid., http://www.pnas.org/content/early
      /2018/02/20/1718503115.
102   Another study found that: E. C. South, B. C. Hohl, M. C. Kondo,
      J. M. MacDonald and C. C. Branas, "Effect of greening vacant land on men-
      tal health of community-dwelling adults: a cluster randomized trial," *JAMA
      Network Open* 1(3) (2018), p. e180298, https://jamanetwork.com/journals
      /jamanetworkopen/fullarticle/2688343.
103   In the United Kingdom, the GOOP: Dr. Michelle Baybutt and Dr. Alan Far-
      rier, *Greener on the Outside: For Prisons. A Guide to Setting Up and Delivering
      a Prison-Based GOOP Project,* University of Central Lancashire, https://www
      .uclan.ac.uk/research/explore/projects/assets/goop-best-practice-guide.pdf.
103   In Halden Prison in Norway: Lilli Fisher, "Prison, nature and social structure,"
      *Terrapin Bright Green,* 12 August 2016, https://www.terrapinbrightgreen.com
      /blog/2016/08/prison-nature-social-structure/.
103   A correctional officer: Nalini Nadkarni, "Life science in prison," TED Talk,

February 2010, https://www.ted.com/talks/nalini_nadkarni_life_science_in
_prison.

104 "I taught the men that mosses": Nalini Nadkarni, "Into the light: bringing
science education to the incarcerated," *Dimensions,* September 2014, https://
www.astc.org/astc-dimensions/into-the-light-bringing-science-education-to
-the-incarcerated/.

104 It was called the Nature Imagery: Nalini Nadkarni, "Nature Imagery in Prisons Project," *Sustainability in Prisons,* 31 August 2017, http://sustainabilityin
prisons.org/blog/2017/08/31/nature-imagery-in-prisons-project/.

104 For the study, inmates: Nalini Nadkarni et al., "Impacts of nature imagery
on people in severely nature-deprived environments," *Frontiers in Ecology
and the Environment* 15(7) (September 2017), pp. 395–403, https://esajournals
.onlinelibrary.wiley.com/doi/abs/10:1002/fee.1518.

105 People who live near parks: In 2003, a study of ten thousand residents in the
Netherlands found that those with a high percentage of green space nearby
(between one and three kilometres) reported better health, including mental
health. Crucially, the study controlled for socio-economic and demographic
factors. For example, wealthy people tend to be healthier and more able to
move out of densely populated areas to green areas if they want, which can skew
results. The research suggested that the elderly, housewives (who spent more
time in their homes) and people in lower socio-economic groups benefited
the most from being near nature. S. de Vries, R. A. Verheij, P. P. Groenewe-
gen and P. Spreeuwenberg, "Natural environments—healthy environments?
An exploratory analysis of the relationship between greenspace and health,"
*Environment and Planning A: Economy and Space* 35(10) (2003), pp. 1717–31,
http://journals.sagepub.com/doi/pdf/10:1068/a35111.

105 The incidence of depression: One of the most significant pieces of work by
the Truro-based team from the European Centre for Environment & Human
Health at the University of Essex was a study of mental health and green space
using data from the British Household Panel. It was the first longitudinal analysis which looked at the health of people who moved between residential areas
with varied amounts of nature, greenery and vegetation. The data followed
the same twelve thousand people over eighteen years as they moved around
England, which gave a much more robust estimate of the effect of nature on
their lives over time. The size of the effects was measured by a comparison
with life events associated with good mental health, such as getting married
or finding a job after a period of unemployment. Moving to an area with a lot
of green space from an area with little green space had a third of the effect of
getting married and a tenth of the effect of becoming employed. The benefits
might be smaller at an individual level, but at population level they could be
substantial, said environmental psychologist Mathew White, one of the lead
researchers. "Imagine you have a park that's visited by 1,000 people," he continued. "Take that park away and build an office block and 100 people get
employed. The utility overall in that community is identical. You've made 100

people a lot happier and 1,000 people a little bit less happy." Mathew P. White et al., "Would you be happier living in a greener urban area? A fixed-effects analysis of panel data," *Psychological Science* 24(6) (April 2013), pp. 920–28, https://journals.sagepub.com/doi/abs/10:1177/0956797612464659.

105   Nature is not a luxury: In 2006, scientists in the Netherlands set out to measure the strength of the relationship between the amount of nature or green space in a person's vicinity, and their health. This time, the study used an even larger data-set and looked at 250,000 residents. The team investigated the differences between urban and rural areas and found that health differences could be partly explained by the presence or lack of nature. "Green space seems to be more than just a luxury and consequently the development of green space should be allocated a more central position in spatial planning policy," concluded the authors. Jolanda Maas et al., "Green space, urbanity, and health: how strong is the relation?," *Journal of Epidemiology & Community Health,* 60(7) (July 2006), pp. 587–92, https://jech.bmj.com/content/60/7/587.

105   There is a direct benefit: In the United States, epidemiologist Peter James, at Harvard University's Department of Population Medicine, conducted a large data-set study. He analysed the biennial health questionnaires of 108,630 nurses living in eleven states in North America between 2000 and 2008 and found that higher levels of greenness equated to decreased mortality. Women in the top fifth of residential greenness in a 250-metre area around their house had a 12 percent lower rate of mortality than those in the lowest fifth, a significant gap. P. James et al., "Exposure to greenness and mortality in a nationwide prospective cohort study of women," *Environmental Health Perspectives* 124(9) (September 2016), https://ehp.niehs.nih.gov/doi/10:1289/ehp.1510363.

To find out what factors might explain the association between greenness and mortality, they collected information on doctor-diagnosed depression and antidepressant medication. They discovered that the depression pathway explained 30 percent of the association between greenness and mortality. Increased greenness was associated with lower depression, which explained the lower mortality rates. "We weren't expecting the magnitude. That 30 percent number was larger than we expected. We expected it to be small," says James. It suggested that Wilson's theory was true: "That there's a direct cognitive benefit and restorative quality of being in nature, that we've evolved in nature to enjoy being in nature," James explains. The study concluded that higher levels of vegetation improve both physical and mental health and recommended that planning policies should include nature in residential areas. "Policies to increase vegetation may provide opportunities for physical activity, reduce harmful exposures, increase social engagement, and improve mental health." James was quick to point out this isn't just about moving to the countryside. With 84 percent of people in the United States living in urban areas, the study suggests that street trees and vegetation within urban areas can have a significant benefit for health.

105 As the impacts of man-made: M. Wolf, "Why climate change puts the poorest most at risk," *Financial Times,* 17 October 2017, https://www.ft.com/content /f350020e-b206-11e7-a398-73d59db9e399.

105 Children, women, the elderly: D. Filiberto et al., "Older people and climate change: vulnerability and health effects," *Generations,* American Society on Aging, Winter 2009–2010, https://www.asaging.org/blog/older-people-and -climate-change-vulnerability-and-health-effects.

106 Friends of the Earth: I. Johnston, "Golf courses cover 10 times more land than allotments—and get £550,000 in farming subsidies," *The Independent,* 21 July 2017, https://www.independent.co.uk/environment/golf-courses-farming -subsidies-allotments-michael-gove-environment-secretary-cap-eu-wealthy -a7853741.html.

106 People in lower socio-economic: "Just 56% of under-16s from BAME house-holds visited the natural environment at least once a week, compared to 74% from white households." A. Leach, "Improving children's access to nature starts with addressing inequality," *The Guardian,* 1 March 2018, https:// www.theguardian.com/teacher-network/2018/mar/01/improving-childrens -access-nature-addressing-inequality-bame-low-income-backgrounds.

106 Evidence suggests that: A government White Paper on the natural envi-ronment found that access to and use of green space are lower among peo-ple who are elderly, have a disability, are from a BAME background and/or live in a deprived area. HM Government, *The Natural Choice: securing the value of nature* (CM8082) (June 2011), https://assets.publishing.service .gov.uk/government/uploads/system/uploads/attachment_data/file/228842 /8082.pdf.); J. Gelsthorpe, *Disconnect from Nature and Its Effect on Health and Well-being: A Public Engagement Literature Review,* Natural History Museum, April 2017, http://www.nhm.ac.uk/content/dam/nhmwww /about-us/visitor-research/Disconnect%20with%20nature%20Lit%20review .pdf.

106 In towns and cities: Gelsthorpe, *Disconnect from Nature and Its Effect on Health and Well-being.*

106 Children who live in deprived: R. Balfour and J. Allen, "Local action on health inequalities: improving access to green spaces," *Public Health En-gland,* September 2014, https://assets.publishing.service.gov.uk/government /uploads/system/uploads/attachment_data/file/357411/Review8_Green_spaces _health_inequalities.pdf.

106 Children living in the poorest: R. Mitchell and F. Popham, "Effect of expo-sure to natural environment on health inequalities: an observational popu-lation study," *The Lancet* 372(9650) (November 2008), pp. 1655–60, https:// www.thelancet.com/journals/lancet/article/PIIS0140-6736(08)61689-X /fulltext.

107 "We ramblers after a hard week's": Benny Rothman, *Green World,* https:// greenworld.org.uk/article/celebrating-1932-kinder-scout-mass-trespass.

108 In an essay for *Outside* magazine: L. Graham, "We're here. You just don't see

us," *Outside,* 1 May 2018, https://www.outsideonline.com/2296351/were-here -you-just-dont-see-us.

108   Soon enough, they become engaged: Personal interview with David Lindo, 22 March 2018, https://theurbanbirder.com/.

108   "Intentional and mindful programming": Personal interview with Howard Frumkin, 21 May 2018.

109   Mitchell's concept is known: Rich Mitchell, "What is equigenesis and how might it help narrow health inequalities?" *Cresh,* 8 November 2013, https:// cresh.org.uk/2013/11/08/what-is-equigenesis-and-how-might-it-help-narrow -health-inequalities/.

109   It became clear to him: Personal interview with Richard Mitchell, 20 September 2017.

110   Their paper was published: Mitchell and Popham, "Effect of exposure to natural environment on health inequalities."

111   Since then, Mitchell and others: R. J. Mitchell, E. A. Richardson, N. K. Shortt and J. R. Pearce, "Neighborhood environments and socioeconomic inequalities in mental well-being," *American Journal of Preventive Medicine* 49(1) (July 2015), pp. 80–84, https://www.ncbi.nlm.nih.gov/pubmed/25911270.

7 · ECOLOGICAL GRIEF

115   Then, he wrote to the "Letters": Michelle Nijhuis, "What do you call the last of a species?" *The New Yorker,* 2 March 2017, https://www.newyorker.com /tech/annals-of-technology/what-do-you-call-the-last-of-a-species.

115   Artists of all kinds were: https://cull.bandcamp.com/album/endling.

115   Instead, they described the situation: G. Ceballos, P. R. Ehrlich and R. Dirzo, "Biological annihilation via the ongoing sixth mass extinction signaled by vertebrate population losses and declines," *Proceedings of the National Academy of Sciences* 114(30) (July 2017), pp. e6089–e6096.

116   The company planned to: J. Dalton, "Councils could be banned from felling trees without consulting locals after three years of Sheffield protests," *The Independent,* 30 December 2018, https://www.independent.co.uk/news /uk/home-news/trees-councils-fell-chop-down-residents-rights-urban-areas -michael-gove-a8703546.html.

116   Wildlife groups mobilized: "Statement on the Chelsea Road elm in Sheffield," *Butterfly Conservation,* 13 February 2018, https://butterfly-conservation.org /news-and-blog/statement-on-the-chelsea-road-elm-in-sheffield.

117   It had an added gravity: Dutch elm disease, https://www.forestresearch.gov .uk/tools-and-resources/pest-and-disease-resources/dutch-elm-disease/.

117   In February 2019, the tree: https://twitter.com/PaulSelby18/status/1096 117152107692032.

117   Thankfully, the Sheffield: "Sheffield tree felling: more saved after deal brokered," *BBC News,* 24 October 2018, https://www.bbc.co.uk/news/uk -england-south-yorkshire-45958502.

117 "In most cases," said Sally: C. Burn, "Sheffield Council blames 'exceptional pressure' in apology for misleading residents over tree-felling work," *Yorkshire Post,* 27 February 2019, https://www.yorkshirepost.co.uk/news/sheffield -council-blames-exceptional-pressure-in-apology-for-misleading-residents -over-tree-felling-work-1-9619446.

117 Elsewhere in Europe: Bogdan Jaroszewiez, Olga Cholewin´ska, Jerzy M. Gutowski et al., "Białowieża Forest—a relic of the high naturalness of European forests," *Forests* 10(10), 849 (2019), https://doi.org/10:3390/f10100849.

117 Of all forests in Europe: A. Nelsen, "Poland violated EU laws by logging in Białowieża forest, court rules," *The Guardian,* 17 April 2018, https://www .theguardian.com/world/2018/apr/17/poland-violated-eu-laws-by-logging-in -biaowieza-forest-says-ecj.

118 They found significant: "The first results on interaction with cancer proliferation genes are promising and may result in development of new anticancer drug candidates," said Zjawiony in a personal email to me.

119 Of the three thousand plants: S. Pandey, P. N. Shaw and A. K. Hewavitharana, "Review of procedures used for the extraction of anti-cancer compounds from tropical plants," *Anti-cancer Agents in Medicinal Chemistry* 15(3) (2015), pp. 314–26, https://www.ncbi.nlm.nih.gov/pubmed/25403166.

119 the deforestation of the Amazon: M. Savarese, "Brazil's Bolsonaro plans to axe environmental panel that protects Amazon rainforest," *The Independent,* 8 April 2019, https://www.independent.co.uk/news/world/americas /bolsonaro-brazil-deforestation-amazon-rainforest-a8860311.html.

120 It was absurd and ridiculous: Personal conversation with Ewa Zin, 3 February 2018.

121 "At some point there will be a collapse": C. Davies, " 'My worst nightmares are coming true': last major primeval forest in Europe on 'brink of collapse.' " *The Guardian,* 23 May 2017, https://www.theguardian.com/environment/2017 /may/23/worst-nightmare-europes-last-primeval-forest-brink-collapse-logging.

121 Estimates of the numbers: "Scars of the primeval forest," http://infografiki .biqdata.pl/wyrok-na-puszcze/en/desktop.html.

121 Plots were logged throughout: Ibid.

121 When I was there: Court of Justice of the European Union, Advocate General's Opinion in Case C-441/17 *Commission v Poland,* No. 13/18, Luxembourg, 20 February 2018, https://curia.europa.eu/jcms/upload/docs/application/pdf /2018-02/cp180013en.pdf.

122 Polls suggest over 80: "Lawyers sound alarm over new Polish plans to log Bialowieza Forest," *Client Earth,* 28 January 2019, https://www.clientearth .org/lawyers-sound-alarm-over-new-polish-plans-to-log-bialowieza-forest/.

122 In April 2018, the ECJ: Ibid.

122 A study of antidepressant: Terry Hartig et al., "Cold summer weather, constrained restoration, and the use of antidepressants in Sweden," *Journal of Environmental Psychology* 27(2) (June 2007), pp. 107–16, https://www .sciencedirect.com/science/article/pii/S0272494407000205.

123 Emerging evidence links: R. N. Salas, P. Knappenberger and J. Hess, *2018 Lancet Countdown on Health and Climate Change Brief for the United States of America,* The Lancet Countdown, London, November 2018.

123 In the Nunatsiavut region: L. Albeck-Ripka, "Why lost ice means lost hope for an Inuit village," *New York Times,* 25 November 2017, https://www.nytimes.com/interactive/2017/11/25/climate/arctic-climate-change.html.

123 What happens in the Arctic: E. Holthaus, "Let it go: the Arctic will never be frozen again," *Grist,* 18 December 2017, https://grist.org/article/let-it-go-the-arctic-will-never-be-frozen-again/.

123 Climate change will affect: H. L. Berry, K. Bowen and T. Kjellstrom, "Climate change and mental health: a causal pathways framework," *International Journal of Public Health* 55 (2010), p. 123, https://link.springer.com/article/10:1007%2Fs00038-009-0112-0.

124 "It's terrible to know": A. Cunsolo and N. R. Ellis, "Ecological grief as a mental health response to climate change-related loss," *Nature Climate Change* 8 (2018), pp. 275–81, https://www.nature.com/articles/s41558-018-0092-2.

124 This effect has been: Ibid.

124 As Glenn Albrecht, the Australian: Albeck-Ripka, "Why lost ice means lost hope for an Inuit village."

124 "We are involved in a kind": Wendell Berry, "Two Minds," in *Essays 1993–2017* (Library of America, 2019).

125 I can relate to the BirthStrikers: E. Hunt, "BirthStrikers: meet the women who refuse to have children until climate change ends," *The Guardian,* 12 May 2019, https://www.theguardian.com/lifeandstyle/2019/mar/12/birthstrikers-meet-the-women-who-refuse-to-have-children-until-climate-change-ends.

125 This is now, as Alexandria: M. Taylor, "Is Alexandria Ocasio-Cortez right to ask if the climate means we should have fewer children?" *The Guardian,* 27 February 2019, https://www.theguardian.com/environment/shortcuts/2019/feb/27/is-alexandria-ocasio-cortez-right-to-ask-if-the-climate-means-we-should-have-fewer-children.

125 When I think about my: R. Andrews, "Here are the countries most likely to survive climate change," *IFLScience,* 28 August 2017, https://www.iflscience.com/environment/countries-likely-survive-climate-change/.

8 · THE FIRST PRIMROSE OF THE YEAR

130 Knowing that every winter will: Alfred, Lord Tennyson, *In Memoriam A.H.H.,* Canto 54.

132 Where will wild swimmers: R. Watts, R. Blakely, G. Greenwood and D. Lewis, "Pollution: no river in England is safe for swimming," *The Times,* 2 August 2019.

132 The symptoms, she wrote: George Sessions, *Deep Ecology for the Twenty-First Century* (Shambhala, 1995), p. 37.

134 "The very thing that is causing": Mary-Jayne Rust, "Climate on the couch:

unconscious processes in relation to our environmental crisis," *Psychotherapy & Politics International* 6(3) (October 2008), pp. 157–70.

135 This made me think of: Nicole Seymour, *Bad Environmentalism: Irony and Irreverence in the Ecological Age* (University of Minnesota Press, 2018), p. 36.

135 "There's all sorts of things": Personal interview with Mary-Jayne Rust, 15 March 2018.

137 As part of one of many: Yew trees fill the churchyards of Britain. It's unclear whether they were there first, or whether the churches were built around the trees, planted by Druids. The painter Caspar David Friedrich used the image of the evergreen fir as a symbol of the resurrection, which Simon Schama describes as "going directly to the heart of one of our most powerful yearnings: the craving to find in nature a consolation for our mortality." Simon Schama, *Landscape and Memory* (HarperCollins, 2004).

138 The woods, he wrote, were: Paul Bishop, *Analytical Psychology and German Classical Aesthetics: Goethe, Schiller, and Jung, Vol. 1: The Development of the Personality* (Routledge, 2007).

138 Writing to a colleague in 1947: *Letters of C. G. Jung, Vol. 1: 1906–1950* (Routledge, 1973), p. 479.

138 The word "numinous": Charlton T. Lewis and Charles Short, *A Latin Dictionary*, http://www.perseus.tufts.edu/hopper/text?doc=Perseus:text:1999 :04.0059.

140 He called it a loss: Meredith Sabini (ed.), *The Earth Has a Soul: C. G. Jung on Nature, Technology & Modern Life* (North Atlantic Books, 2002), pp. 80, 81.

140 Little did he know: M. Taylor, "Climate change making storms like Idai more severe, say experts," *The Guardian,* 19 March 2019, https://www.theguardian .com/world/2019/mar/19/climate-change-making-storms-like-idai-more-severe -say-experts.

140 extreme heat and drought: C. Harvey, "How climate change may affect winter 'weather whiplash,'" *Scientific American,* 11 February 2019, https:// www.scientificamerican.com/article/how-climate-change-may-affect-winter -weather-whiplash/.

140 In 1956, a few years: *Letters of C. G. Jung, Vol. 1: 1906–1950,* p. 320.

## 9 · AND IN THE END . . .

142 In the early twenty-first: S. L. Koole and A. E. Van den Berg, "Lost in the wilderness: terror management, action orientation, and nature evaluation," *Journal of Personality and Social Psychology* 88(6) (June 2005), pp. 1014–28, https:// pdfs.semanticscholar.org/6375/a6626be63945ea8eac817a18c4cdc42bb71b.pdf.

143 Perhaps, as Richard Mabey has: Richard Mabey, *Nature Cure* (Vintage, 2008), p. 223.

143 An experimental psychology: J. L. Goldenberg et al., "I am *not* an animal: mortality salience, disgust, and the denial of human creatureliness," *Journal of Experimental Psychology: General* 130(3) (September 2001), pp. 427–35,

https://www.researchgate.net/publication/232450313_I_Am_Not_an
_Animal_Mortality_Salience_Disgust_and_the_Denial_of_Human
_Creatureliness.

143 The more aware people: Study 1 revealed that reminders of death led to an increased emotional reaction of disgust to body products and animals. Study 2 showed that, compared to a control condition, mortality salience led to greater preference for an essay describing people as distinct from animals.

143 Is our treatment of the elderly: NHS, "Loneliness in older people," https://www .nhs.uk/conditions/stress-anxiety-depression/loneliness-in-older-people/.

143 Suppose, as the American: Loren C. Eiseley, *The Lost Notebooks of Loren Eiseley* (Bison, 2002), p. 115.

145 Entries often go like this: All quotations are from Derek Jarman, *Modern Nature: The Journals of Derek Jarman* (Century, 1991).

## 10 · FUTURE NATURE

151 Some natural philosophers: Johann Wolfgang von Goethe, *Theory of Colours*, translated by Charles Lock Eastlake (John Murray, 1840), p. 33.

151 At the water's edge: https://svalbardmuseum.no/en/kultur-og-historie/hval fangst/.

152 There are currently hundreds: At the time of writing, the number of samples was 984,547.

153 Longyearbyen is now: I. Hanssen-Bauer, E. J. Førland, H. Hisdal, S. Mayer, A. B. Sandø and A. Sorteberg, *Climate in Svalbard 2100*, January 2019, https:// www.miljodirektoratet.no/globalassets/publikasjoner/M1242/M1242.pdf; S. Lazarus, "'Doomsday vault' town warming faster than any other on Earth," CNN, 27 March 2019, https://edition.cnn.com/2019/03/26/europe /longyearbyen-doomsday-vault-climate-change-intl/index.html.

154 In light of the fact: Rainforest Action Network, "How many trees are cut down every year?" *The Understory,* 6 March 2017, https://www.ran.org /theunderstory/how_many_trees_are_cut_down_every_year/.

154 estimates that there will be: T. W. Crowther, H. B. Glick, K. R. Covey et al., "Mapping tree density at a global scale," *Nature* 525 (7568) (10 September 2015), pp. 201–5, https://www.nature.com/articles/nature14967.

154 Precision medicine is being: L. Jones, "Strength in numbers: ultra-tailored, personalised medicine is the grand promise for the future of healthcare," *The Long + Short,* 28 March 2017, https://thelongandshort.org/life-death /precision-medicine.

156 "People did not talk": See Peter H. Kahn Jr., "Technological nature and human well-being," in Matilda van den Bosch and William Bird (eds.), *Oxford Textbook of Nature and Public Health* (Oxford University Press, 2018), Chapter 5:3.

157 "Maybe it's time to get": "[EI] Reading 3: Telegarden (1995) by Ken Goldberg,"

https://oss.adm.ntu.edu.sg/fni0001/ei-reading-3-telegarden-1995-by-ken
-goldberg/.

158 As of 2018, in Northern: https://www.un.org/development/desa/en/news
/population/2018-revision-of-world-urbanization-prospects.html.

159 "These are spaces of conviviality": Devita Davison, "How urban agriculture
is transforming Detroit," TED Talk, 7 December 2017, https://archive.org
/details/DevitaDavison2017.

159 "Every day, we collectively": https://twitter.com/DevitaDavison/status
/1008398879803535363.

160 It also consistently beats: "Healing through Nature: Khoo Teck Puat Hospital," International Living Future Institute, 2017, https://living-future.org
/biophilic/case-studies/award-winner-khoo-teck-puat-hospital/.

160 In 2017, it won: K. Edelstein, "Singapore's Khoo Teck Puat wins first Kellert Biophilic Award," *The Kendeda Fund,* 13 November 2017, https://liv
ingbuilding.kendedafund.org/2017/11/13/singapores-khoo-teck-paut-wins
-first-kellert-biophilic-award/.

160 We have known empirically: R. S. Ulrich, "View through a window may
influence recovery from surgery," *Science* 224 (4647) (April 1984), pp. 420–21,
https://www.ncbi.nlm.nih.gov/pubmed/6143402.

161 Grass is in fact: R. Harrington, "America's biggest crop is not what you
think," *Business Insider,* 19 February 2016, https://www.businessinsider.com
/americas-biggest-crop-is-grass-2016-2?r=US&IR=T.

161 But grass supports little: C. Foran, "California residents are painting their
lawns green," *The Atlantic,* 5 August 2014, https://www.theatlantic.com
/politics/archive/2014/08/california-residents-are-painting-their-lawns-green
/444909/.

161 United Kingdom–based scientist Lionel Smith: Lionel Smith, *Tapestry
Lawns Freed from Grass and Full of Flowers* (Routledge, 2019), https://www
.routledge.com/:%20Tapestry-Lawns-Freed-from-Grass-and-Full-of-Flowers
-1st-Edition/Smith/p/book/9780367144036.

161 In the United States, in Milwaukee: L. Mazur, "Milwaukee is showing how
urban gardening can heal a city," *Civil Eats,* 4 October 2017, https://civileats
.com/2017/10/04/milwaukee-is-showing-how-urban-gardening-can-heal-a-city/.

161 and San Francisco: Urban Orchards, San Francisco Department of the
Environment, https://sfenvironment.org/article/managing-our-urban-forest
-types-of-urban-agriculture/urban-orchards.

161 In Australia, Melbourne: Anne Jaluzot, "Melbourne's coordinated approach
to streetscape projects to double canopy," Trees & Design Action Group,
2014, http://www.tdag.org.uk/uploads/4/2/8/0/4280686/melbourne.pdf.

161 The City's government has: https://www.melbourne.vic.gov.au/community
/parks-open-spaces/urban-forest/Pages/urban-forest-strategy.aspx.

161 Acoustic ecologist Gordon Hempton: https://www.soundtracker.com/about
-gordon-hempton/.

161 Listening to a recording: https://onesquareinch.org/breathing-space/.

162 According to research by: N. Duvergne Smith, "Treepedia: calculating the value of green," *Slice of MIT*, 3 March 2017, https://alum.mit.edu/slice /treepedia-calculating-value-green.

164 In 2014, New Zealand passed: J. Ruru, "Tūhoe-Crown settlement—Te Urewera Act 2014," *Māori Law Review,* October 2014, http://maorilawreview.co .nz/2014/10/tuhoe-crown-settlement-te-urewera-act-2014/.

164 The landscape owns itself: Ibid.

164 This means that the land: "New Zealand," Earth Law Center, 16 August 2016, https://www.earthlawcenter.org/international-law/2016/8/new-zealand.

164 The 2018/2019 Annual Plan: "Bringing the Blueprint to Life, 2018–2019," Annual Plan, Te Uru Taumatua, http://www.ngaituhoe.iwi.nz/annual-plan -bringing-the-blueprint-to-life.

164 The Whanganui River: E. Ainge Roy, "New Zealand river granted same legal rights as human being," *The Guardian,* 16 May 2017, https://www .theguardian.com/world/2017/mar/16/new-zealand-river-granted-same-legal -rights-as-human-being.

164 In the Republic of Benin: https://sacrednaturalsites.org/wp-content/uploads /2014/09/Benin-Sacred-Forest-law-final-English-version-2014.pdf.

164 In India, the heavily: M. Safi, "Ganges and Yamuna rivers granted same legal rights as human beings," *The Guardian,* 21 March 2017, https://www .theguardian.com/world/2017/mar/21/ganges-and-yamuna-rivers-granted -same-legal-rights-as-human-beings.

164 However, the order was: "India's Ganges and Yamuna rivers are 'not living entities,'" *BBC News,* 7 July 2017, https://www.bbc.co.uk/news/world-asia -india-40537701.

164 "When you look at environmental": Personal interview with Veneta Cooney, 30 May 2018.

165 In one document: David Birkbeck and Stefan Kruczkowski, *Building for Life 12,* Building for Life Partnership (Cabe at Design Council, Design for Homes and Home Builders' Federation, 2015) with Nottingham Trent University; and *Homes England Strategic Plan 2018 to 2023,* October 2018, https://www.gov.uk/government/publications/homes-england-strategic-plan -201819-to-202223/homes-england-strategic-plan-2018-to-2023#who-we-are. See also National Planning Policy Framework, March 2012, https://web archive.nationalarchives.gov.uk/20180608095821/https://www.gov.uk/gov ernment/publications/national-planning-policy-framework–2.

### CONCLUSION · A NEW DYAD

169 The trajectory of destruction: Anthony Huxley, *Plant and Planet* (Allen Lane, 1974), p. 2; quoting E. J. H. Corner.

170 "Becoming indigenous to place": Robin Wall Kimmerer, *Braiding Sweetgrass:*

*Indigenous Wisdom, Scientific Knowledge and the Teachings of Plants* (Milk-weed Editions, 2013), p. 205.

171   The reintroduction of species: "Beaver reintroduction in the UK," RSPB, https://www.rspb.org.uk/our-work/our-positions-and-casework/our-positions/species/beaver-reintroduction-in-the-uk/.

171   water voles: "Restoring Ratty," Northumberland Wildlife Trusts, https://www.nwt.org.uk/what-we-do/projects/restoring-ratty.

171   ladybird spiders, bitterns, avocets: S. Tregaskis, "UK animals back from the brink of extinction," *The Guardian,* 18 August 2011, https://www.theguardian.com/environment/blog/2011/aug/18/uk-animals-back-brink-extinction.

171   and otters: P. Barkham, "Otters are back—in every county in England," *The Guardian,* 18 August 2011, https://www.theguardian.com/environment/2011/aug/18/otters-return-british-rivers.

# Bibliography

Abram, David, *The Spell of the Sensuous: Perception and Language in a More-Than-Human World* (Vintage, 1997)

Albrecht, Glenn, *Earth Emotions* (Cornell University Press, 2019)

Ammons, A. R., *The Selected Poems 1951–1977* (Princeton University Press, 1977)

Berry, Thomas, *The Dream of the Earth* (1988; Counterpoint, 2015)

Berry, Wendell, *The World-Ending Fire: The Essential Wendell Berry* (Allen Lane, 2017)

Bullmore, Edward, *The Inflamed Mind: A Radical New Approach to Depression* (Short Books Ltd., 2018)

Buzzell, Linda, and Craig Chalquist (eds.), *Ecotherapy: Healing with Nature in Mind* (Counterpoint, 2010)

Carson, Rachel, *The Sense of Wonder: A Celebration of Nature for Parents and Children* (1965; HarperCollins Publishers, Inc., reprint edition, 2017)

Cobb, Edith, *The Ecology of Imagination in Childhood* (Spring Publications, 1994)

Deakin, Roger, *Waterlog: A Swimmer's Journey Through Britain* (Vintage, 2000)

Francis, Vievee, *Forest Primeval* (Northwestern University Press, 2015)

Griffiths, Jay, *Kith: The Riddle of the Childscape* (Penguin Books, 2014)

———, *Wild: An Elemental Journey* (Penguin Books, 2006)

Halpern, David, *Mental Health and the Built Environment: More than Bricks and Mortar?* (Taylor & Francis, 1995)

Huxley, Anthony, *Plant and Planet* (Allen Lane, 1974)

Jarman, Derek, *Modern Nature: The Journals of Derek Jarman* (1991; Vintage, 2018)

Jones, Kathleen, *Asylums and After: A Revised History of the Mental Health Services: From the Early 18th Century to the 1990s* (Athlone Press, 1993)

Kaplan, Rachel, and Stephen Kaplan, *The Experience of Nature: A Psychological Perspective* (Cambridge University Press, 1989)

Kellert, Stephen R., and Edward O. Wilson (eds.), *The Biophilia Hypothesis* (Island Press, 1993)

Kimmerer, Robin Wall, *Braiding Sweetgrass: Indigenous Wisdom, Scientific Knowledge and the Teachings of Plants* (Milkweed Editions, 2013)

Lewis, Marc, *The Biology of Desire: Why Addiction Is Not a Disease* (Perseus Books, 2015)

Lewis-Stempel, John, *Where Poppies Blow: The British Soldier, Nature, The Great War* (Weidenfeld &Nicolson, 2016)

Li, Dr. Qing, *Shinrin-Yoku: The Art and Science of Forest-Bathing: How Trees Can Help You Find Health and Happiness* (Penguin Books, 2018)

Louv, Richard, *Last Child in the Woods* (Algonquin Books of Chapel Hill, 2008)

Mabey, Richard, *Nature Cure* (Vintage, 2008)

Mellor, Mary, *Feminism and Ecology* (Polity Press, 1997)

Nash, Roderick, *Wilderness and the American Mind* (Yale University Press, 1982)

Orians, Gordon H., *Snakes, Sunrises, and Shakespeare* (University of Chicago Press, 2014)

Rousseau, Jean-Jacques, *Emile* (1762; Dover Publications, 2013)

Sabini, Meredith (ed.), *The Earth Has a Soul: C. G. Jung on Nature, Technology & Modern Life* (North Atlantic Books, 2002)

Schama, Simon, *Landscape and Memory* (HarperCollins Publishers, 2004)

Schulze, Robin G., *The Degenerate Muse: American Nature, Modernist Poetry, and the Problem of Cultural Hygiene* (Oxford University Press, 2014)

Sessions, George (ed.), *Deep Ecology for the Twenty-First Century* (Shambhala, 1995)

Seymour, Nicole, *Bad Environmentalism: Irony and Irreverence in the Ecological Age* (University of Minnesota Press, 2018)

Steg, Linda, Agnes E. van den Berg and Judith I. M. de Groot (eds.), *Environmental Psychology: An Introduction* (Wiley-Blackwell, 2012)

Thomas, Edward, *Selected Poems* (Faber & Faber, 2011)

Thomas, Keith, *Man and the Natural World: Changing Attitudes in England 1500– 1800* (1983; Penguin Books, new edition, 1991)

Thoreau, Henry David, *Walden* (1854; Penguin Books, 2016)

Tsing, Anna Lowenhaupt, *The Mushroom at the End of the World: On the Possibility of Life in Capitalist Ruins* (Princeton University Press, 2015)

Van den Bosch, Matilda, and William Bird (eds.), *Oxford Textbook of Nature and Public Health* (Oxford University Press, 2018)

Wilson, Edward O., *Biophilia* (Harvard University Press, 1984)

# Index

Printed in the United States
by Baker & Taylor Publisher Services